T0238293

Lecture Notes in Computer Science   1729

Edited by G. Goos, J. Hartmanis and J. van Leeuwen

Springer

Berlin
Heidelberg
New York
Barcelona
Hong Kong
London
Milan
Paris
Singapore
Tokyo

Masahiro Mambo   Yuliang Zheng (Eds.)

Information Security

Second International Workshop, ISW'99
Kuala Lumpur, Malaysia, November 6-7, 1999
Proceedings

Springer

Series Editors

Gerhard Goos, Karlsruhe University, Germany
Juris Hartmanis, Cornell University, NY, USA
Jan van Leeuwen, Utrecht University, The Netherlands

Volume Editors

Masahiro Mambo
Tohoku University, Education Center for Information Processing
Kawauchi Aoba Sendai, 980-8576, Japan
E-mail: mambo@ecip.tohoku.ac.jp

Yuliang Zheng
Monash University, School of Computer and Information Technology
MacMahons Road, Frankston, Melbourne, Victoria 3199, Australia
E-mail: yuliang.zheng@infotech.monash.edu.au

Cataloging-in-Publication data applied for

Die Deutsche Bibliothek - CIP-Einheitsaufnahme

Information security : second international workshop ; proceedings /
ISW '99, Kuala Lumpur, Malaysia, November 6 - 7, 1999. Masahiro
Mambo ; Yuliang Zheng (ed.). - Berlin ; Heidelberg ; New York ;
Barcelona ; Hong Kong ; London ; Milan ; Paris ; Singapore ; Tokyo
: Springer, 1999
  (Lecture notes in computer science ; Vol. 1729)
  ISBN 3-540-66695-8

CR Subject Classification (1998): E.3, G.2.1, D.4.6, K.6.5, F.2.1-2, C.2, J.1

ISSN 0302-9743
ISBN 3-540-66695-8 Springer-Verlag Berlin Heidelberg New York

Typesetting: Camera-ready by author
SPIN: 10703333   06/3142 – 5 4 3 2 1 0   Printed on acid-free paper

Preface

The 1999 International Information Security Workshop, ISW'99, was held on Monash University's Malaysia Campus, which is about 20km to the south west of downtown Kuala Lumpur, November 6-7, 1999.

ISW'99 sought a different goal from its predecessor, ISW'97, held in Ishikawa, Japan, whose proceedings were published as Volume 1396 of Springer Verlag's LNCS series. The focus of ISW'99 was on the following emerging areas of importance in information security: multimedia watermarking, electronic cash, secure software components and mobile agents, and protection of software.

The program committee received 38 full submissions from 12 countries and regions: Australia, China, France, Germany, Hong Kong, Japan, Korea, Malaysia, Singapore, Spain, Taiwan, and USA, and selected 23 of them for presentation. Among the 23 presentations, 19 were regular talks and the remaining 4 were short talks. Each submission was reviewed by at least two expert referees.

We are grateful to the members of the program committee for reviewing and selecting papers in a very short period of time. Their comments helped the authors improve the final version of their papers. Our thanks also go to Patrick McDaniel, Masaji Kawahara, and Yasuhiro Ohtaki who assisted in reviewing papers. In addition, we would like to thank all the authors, including those whose submissions were not accepted, for their contribution to the success of this workshop.

The workshop was organized with the help of local committee members, including Cheang Kok Soon, Hiew Pang Leang, Lily Leong, and Robin Pollard. We appreciate their patience and professionalism. Robin Pollard led the committee as a general co-chair. We owe the success of the workshop to him as well as general co-chair Eiji Okamoto.

November 1999

Masahiro Mambo
Yuliang Zheng

Information Security Workshop (ISW'99)

Organizing Committee

Eiji Okamoto (**Co-chair**, Univ. of Wisconsin, Milwaukee, USA)
Robin Pollard (**Co-chair**, Monash University, Malaysia)
Hiew Pang Leang (Monash University, Malaysia)
Lily Leong (Monash University, Malaysia)
Cheang Kok Soon (Monash University, Malaysia)

Program Committee

Masahiro Mambo, (**Co-chair**, Tohoku University, Japan)
Yuliang Zheng, (**Co-chair**, Monash University, Australia)
David Aucsmith (Intel, USA)
George Davida (University of Wisconsin-Milwaukee, USA)
Robert H. Deng (Kent Ridge Digital Labs, Singapore)
Steven J. Greenwald (Independent Consultant, USA)
Ryoichi Mori (Superdistribution Laboratory, Japan)
Kazuo Ohta (NTT, Japan)
Aviel Rubin (AT&T Labs - Research, USA)
Andrew Z Tirkel (Monash University, Australia)
Moti Yung (CertCo, USA)

Contents

Electronic Money

Electronic Payment and Unlinkability

Secure Software Components, Mobile Agents, and Authentication

Network Security

Digital Watermarking

Protection of Software and Data

Electronic Money, Key Recovery, and Electronic Voting

Digital Signatures

Spending Programs: A Tool for Flexible Micropayments*

Josep Domingo-Ferrer and Jordi Herrera-Joancomartí

Universitat Rovira i Virgili, Department of Computer Science and Mathematics,
Autovia de Salou s/n, E-43006 Tarragona, Catalonia, Spain,
{jdomingo,jherrera}@etse.urv.es

Abstract. Micropayments are electronic payments of small amount.
Given their low value, the cost of the corresponding electronic transactions should also be kept low. Current micropayment schemes allow a
regular amount of money withdrawn from a bank to be split into fixed-
value coupons, each of which is used for one micropayment. A more
flexible mechanism is proposed in this paper, whereby coupons of variable value can be generated by a *spending program* without increasing
the transaction cost. Moreover, the spending program allows one of several alternative ways of splitting the amount withdrawn into re-usable
coupons to be selected in real-time.

Keywords: Micropayments, Electronic commerce, Hash functions, Hash
chain, Spending program.

1 Introduction

Micropayments are electronic payments of low value and they are called to playing a major role in the expansion of electronic commerce: example applications
are phone call payments, access to non-free web pages, pay-per-view TV, etc.
The reason for designing specific micropayment schemes is that standard electronic payment systems (like CyberCash [3], e-cash [6] , *i*KP [2], SET [16]) for
low-value payments suffer from too high transaction costs as compared to the
amount of payments. The cost of transactions is kept high due to complex cryptographic protocols like digital signatures used for achieving a certain security
level. However, micropayments do not need as much security as speed and simplicity (in terms of computing). For that reason, several micropayment proposals
try to replace the use of digital signatures with faster operations.

1.1 Our Result

Current micropayment schemes allow a regular amount of money withdrawn
from a bank to be split into fixed-value coupons, each of which is used for one
micropayment. A more flexible mechanism is proposed in this paper, whereby

---

* This work is partly supported by the Spanish CICYT under grant no. TEL98-0699-
C02-02.

M. Mambo, Y. Zheng (Eds.): ISW'99, LNCS 1729, pp. 1–13, 1999.

coupons of variable value can be generated, several currencies can be used in successive micropayments, and larger payments can be made without computational overcost for the merchant or the buyer. Moreover, the buyer can provide input at transaction time to select, skip or re-use coupons.

1.2   Plan of This Paper

Section 2 contains some background on hash-based micropayment schemes. Section 3 introduces the concept of spending program. Section 4 discusses non-iterative spending programs (where coupons cannot be re-used). Section 5 presents iterative spending programs (where coupons are re-used). Section 6 is a conclusion. The Appendix recalls the structural program coding which is used to ensure integrity for spending programs.

2   Background on Hash-Based Micropayments

Quite a number of micropayment systems can be found in the literature that use hash functions instead of digital signatures to reduce the computational burden. On a typical workstation, it may take half a second to compute an RSA [15] signature; in that period, 100 RSA signatures can be verified (assuming a small public exponent) and, more important, 10000 hash functions can be computed. Thus, unlike digital signatures, hash functions allow high-rate verification of micropayments by the merchant without committing too many computing resources. This is a key issue since, for low-value payments to be profitable, they must be collected in an inexpensive way and possibly on a large scale (*i.e.* from a large community of buyers). On the buyer's side, replacing digital signatures with hash functions facilitates the use of smart cards, which are very convenient portable devices but have little computing power. So the advantages of dropping digital signatures in favour of hash functions should be clear.

Micropayment systems based on hash functions include NetCard [1], μ-iKP [8] and PayWord [14]. The principle behind those systems is similar. Let F be a computationally secure one-way hash function (*i.e.* easy to compute and hard to invert). Now the buyer takes a value X that will be the root of the chain and computes the sequence $T_n, T_{n-1}, \cdots, T_0$, where

$$T_0 = F(T_1)$$
$$T_1 = F(T_2)$$
$$\vdots$$
$$T_{n-1} = F(T_n)$$
$$T_n = X$$

(1)

The values $T_1, \cdots, T_n$ are called coupons and will be used by the buyer to perform n micropayments to the same merchant. Each coupon has the same fixed value v. Before the first micropayment, the buyer sends T_0 to the merchant together with the value v in an authenticated manner. The micropayments are thereafter made by successively revealing $T_1, \cdots, T_n$ to the merchant, who can check the validity of T_i by just verifying that $F(T_i) = T_{i-1}$.

We next mention some differences between the main micropayment systems based on hash functions.

NetCard and μ-iKP are both micropayment schemes bootstrapped with normal e-payment systems, SET and iKP:

- With NetCard the bank supplies the root X of the hash chain to the buyer. The buyer then computes the chain, signs its last element T_0, the total number of elements n and the value of each chain element v. These signed values are sent by the buyer to the merchant, who uses the SET protocol to obtain on-line authorization for the whole chain.
- With μ-iKP, the root of the chain is a random value chosen by the buyer and the payment structure is the same as in the iKP payment system. In other words, the on-line authorization of the chain is performed by authorizing a single iKP payment of regular amount.

PayWord [14] is a credit-based scheme that needs a broker. The buyer establishes an account with the broker who gives her a certificate that contains the buyer identity, the broker identity, the public key of the buyer, an expiration date and some other information. The hash chain is produced by the buyer using a random root. When the buyer wants to make a purchase, she sends to the merchant a commitment to a chain. The commitment includes the merchant's identity, the broker certificate, the last element of the chain, the current date, the length of the chain and some other information. In this scheme, the broker certificate certifies that the broker will redeem any payment that the buyer makes before the expiration date, and the buyer commitment authorizes the broker to pay the merchant. Notice that in this scheme the chain is related to a pair buyer/merchant through the commitment. After that, micropayments are made by the buyer by revealing successive elements of the chain to the merchant. PayTree [9] is an extension to PayWord which uses hash trees rather than hash chains; a hash tree is a tree whose leaf nodes are labeled by secret random values, whose internal nodes are labeled by the hash value of the nodes' successors, and whose root is signed. PayTree allows the buyer to use parts of the hash tree to pay multiple merchants, with possibly several different denominations or currencies.

Pedersen [12] also iterates a hash function with a random root to obtain a chain of coins but he does not provide much detail on what kind of system (credit or debit based) he implements nor does he give information about some other security issues.

The authors of μ-iKP emphasize that the use of hash chains implicitly assumes that micropayments take place repeatedly from the same buyer to the same merchant. Such stability assumption on buyer-merchant relationship can be relaxed at the cost of trusting an intermediate broker who maintains stable relationships with several buyers and several merchants: a buyer can send coupons to the broker and the broker is trusted to relay (his own) coupons to the merchant for the same value.

3   Basic Construction

With the exception of PayTree, the micropayment systems described in Section 2 share a lack of flexibility, which results in at least two shortcomings:

- Since all coupons have the same value v, the only way to be able to pay any amount is to let v be the minimal value, for instance one cent. But this means that the merchant must verify fifty hash functions to get paid a sum as small as fifty cents (being undesirable, note that this is still faster than verifying one RSA signature!, see Section 2). It is true that the buyer just needs to send one hash value (the 50th), but in any case she must store or compute all intermediate hashes.
- Fixed-value coupons do not allow to deal with different currencies.

PayTree mitigates the above lack of flexibility by replacing hash chains with hash trees, but still does not allow coupons to be re-used or dynamically selected. The scheme presented in this paper goes one step further and uses a structure more general than a hash tree, namely a *spending program*:

Definition 1 (Spending program). *A spending program $i_1, \cdots, i_n$ is a program whose instructions i_k are either value instructions, flow-control instructions, input-output instructions or assignment instructions.*

Definition 2 (Value instruction). *A value instruction is one that carries a specific sum of money in a currency specified in the same instruction. When a value instruction of a spending program is retrieved by the merchant, the corresponding sum of money is spent by the buyer.*

Definition 3 (Flow-control instruction). *A flow-control instruction allows to modify the flow of a spending program. Four types of flow-control instructions are used:*

1. *Forward unconditional branch*
2. *Forward conditional branch*
3. *Backward unconditional branch*
4. *Backward conditional branch*

If i_k is a branch to instruction i_j, "forward" means that $k < j$ and "backward" that $k \geq j$. Backward branches allow instruction blocks to be executed more than once.

Input-output and assignment instructions are analogous to machine language instructions of the same type.

Clearly, a spending program generalizes the hash chain concept implemented by equations (1) and the hash tree concept used in PayTree, since the value instructions are in fact coupons of arbitrary value and iterations are allowed. An essential issue is to find a way to encode spending programs such that value instructions in an encoded spending program are as unforgeable as coupons in a hash-chain.

The hash-chain idea was first published in [11] applied to password authentication. In [4][5] a generalization of hash chains called structural coding was applied to the program integrity problem. Structural coding and its properties are recalled in the Appendix (the coding is called structural because it depends on the flow-control structure of the program). Thus, both hash chains and structural coding were invented well before hash-based micropayment systems and for quite different purposes. However, just like hash chains turned out to be natural to implement fixed-value coupons, structural coding is a natural tool to implement spending programs:

Protocol 1 (Spending program micropayment)

1. **[Writing]** *The buyer writes the (unencoded) spending program $i_1, \cdots, i_n$ following her own taste. The buyer's computer or smart card can be used to edit the spending program. Alternatively, "ready-made" standard spending programs supplied by the buyer's bank or other institutions can be used.*

2. **[Coding]** *Spending program instructions $i_1, \cdots, i_n$ are encoded by the buyer into a sequence of so-called traces $T_0, T_1, \cdots, T_n$ using the structural coding [5] based on a one-way hash function F such that in general $F(X \oplus Y) \neq F(X) \oplus F(Y)$, where $\oplus$ denotes bitwise exclusive OR (MD5 [13] or SHA [17] are good candidates for F). Coding could also be performed by the buyer's computer or smart card.*

3. **[Signature]** *The first trace T_0 is signed by the buyer or the buyer's card; let t_0 be the signed form of T_0. The buyer also signs an upper bound α_{t_0} of the amount that can be spent using the spending program that starts with T_0.*

4. **[Initialization]** *The buyer sends t_0 and the signed α_{t_0} to the merchant, who uses a standard payment protocol to obtain on-line authorization for the whole spending program. Note that authorizing an upper bound of the spendable amount does not mean that the buyer is paying anything; payment will be done when value instructions of the spending program are sent by the buyer to the merchant.*

5. **[Microspending]** *Before executing an instruction i_k, it must be decoded from the corresponding trace T_k. Decoding is only successful if no modifications have been done to the program in the traces preceding T_k (run-time integrity property, see Appendix). In [5], a preprocessor for decoding traces was assumed to be pipe-lined and encapsulated with the main processor executing instructions. For spending programs, the procedure is slightly different:*

    (a) *When a value instruction (coupon) is to be paid by the buyer to the merchant, the corresponding trace is sent by the buyer to the merchant, maybe after some flow-control and input-output intermediate traces.*

    (b) *The merchant decodes the instructions in the traces received, to check whether these are valid traces. In particular, if the trace corresponding to the value instruction is correctly decoded, then payment for that value is accepted by the merchant.*

  6. [**Clearing**] *The merchant relays the traces received (either one by one or in batches to save transaction costs) to his bank. The bank decodes traces the same way the merchant did, to ensure they are valid (the merchant could try to forge non-existing micropayments). For each correctly decoded instruction value, the bank credits the merchant the corresponding value.*

Note 4. The standard payment system referred to at Step 4 could be for example SET or iKP. For instance, to use iKP, t_0 and α_{t_0} would be placed in the COMMON field defined by that protocol (this is a field containing information shared by all parties involved in the payment, see [2] for more details). This approach is the same adopted by NetCard and μ-iKP, which rely on SET, respectively iKP, for authorization of hash chains.

4 Non-iterative Spending Programs

Since spending programs contain value instructions than can be accepted as payments, an important distinction is whether instructions can be executed more than once.

Definition 5 (Non-iterative spending program). *A spending program i_1, $\cdots$, i_n is called non-iterative if it contains no backward branches, i.e. if no instruction can be executed more than once.*

    Thus in a non-iterative spending program, value instructions cannot be reused. This simplifies matters, because the bare knowledge by the merchant of a trace encoding a value instruction is sufficient for the bank to credit the merchant the corresponding value. We next give an example of sequential spending program with value instructions for several different denominations and/or currencies.

Example 6. The structural coding corresponding to a sequential block of n value instructions with values $v_1, v_2, \cdots, v_n$ is next given (sequential block means that no flow-control instructions are present). Let V_i be the amount v_i with some redundancy (in fact the currency name can be used as redundancy). Let F be a one-way hash function as specified in Section 3 and let $\oplus$ denote bitwise exclusive OR. Then the buyer computes the trace sequence $T_n, T_{n-1}, \cdots, T_1, T_0$ as follows:

$$T_0 = F(T_1) \oplus V_1$$
$$T_1 = F(T_2) \oplus V_2$$

$$\vdots \qquad\qquad (2)$$
$$T_{n-1} = F(T_n) \oplus V_n$$
$$T_n = F(V_n)$$

After the above computation, the buyer signs T_0 to get t_0. Before the first micropayment, the buyer sends t_0 and the signed value $\alpha_{t_0} = \sum_{i=1}^{n} v_i$ to the merchant for authorization. Micropayments of values $v_1, v_2, \cdots, v_n$ can be then made by successively revealing $T_1, \cdots, T_n$ to the merchant. For instance, when the merchant gets T_1, then he can retrieve v_1 by computing $F(T_1) \oplus T_0 = V_1$. The same check is performed by the merchant's bank before crediting V_1 to the merchant's account. ◇

The same structure presented in Example 6 could accomodate multicurrency micropayments. For example, v_1 could be in euros, v_2 in dollars, etc. Of course, with a sequential block, this means that euros should be spent first and dollars second, which is rather rigid. Anyway, current micropayment systems offering only fixed-value coupons are worse in that they must rely on an intermediate broker to deal with multiple currencies; this has at least two drawbacks:

- The broker must be trusted
- The broker cannot be expected to do the currency conversion for free

A solution to increase the flexibility offered by the spending program of Example 6 is to use input-output and forward flow-control instructions (forward branches). In this way, *the buyer could provide input at transaction-time which would be used to select some value instructions and skip some others*. Note that such dynamical selection frees the buyer from knowing before signing the first trace which value instructions he will use. For example, if multicurrency value instructions are being used, the selected currency would be a transaction-time input. Note also that skipping value instructions does not mean losing money because money is only spent when the merchant retrieves a value instruction. See the Appendix on how to encode forward branches while ensuring instruction unforgeability.

A second advantage of forward flow-control is that we can include in the spending program value instructions for large sums (which can be optionally used instead of "micro-value" instructions). In this way, the distinction between micropayments and standard payments becomes rather fuzzy.

5 Iterative Spending Programs

Iterative spending programs are programs in which some instructions can be executed more than once.

Definition 7 (Iterative spending program). *A spending program $i_1, \cdots, i_n$ is called iterative if it contains at least a backward branch, i.e. if some instructions can be executed more than once.*

Clearly, iterative spending programs add new flexibility advantages:

Storage reduction For example, an arbitrary number of value instructions of equal value v can be stored as a two or three instruction loop using a backward branch.

Complex spending patterns Case-like structures can be created in a spending program so that value instructions can be repeatedly selected or skipped as a function of the buyer's transaction-time input.

Just as clearly, iterative spending programs pose a new security problem. Since value instructions can be re-used, once the merchant has completed a whole loop from an iterative spending program, he could present the value instructions in that loop to his bank an arbitrary number of times pretending to be credited more than he is entitled to.

A way to repair the above security flaw is to require that some additional information should accompany the value instructions in a loop if they are used more than once. Specifically, we propose that a Kerberos-like ticket [10] be sent by the buyer to the merchant along with a re-used value instruction. The following protocol can be run in parallel to Protocol 1:

Protocol 2 (Ticket usage)

1. [**Initialization**] *To use Kerberos tickets, the buyer and the merchant's bank must share a symmetric encryption key K. Such key can be agreed upon by the buyer and the merchant's bank using their public-key certificates at Step 4 of Protocol 1 (initialization). It should be noted that, since micropayment assumes stability in buyer-merchant relationship, the overhead of a key exchange between the buyer and the merchant's bank should be affordable.*

2. [**Microspending**] *Each time a value instruction is re-used (and this happens for instructions in a loop always but the first time), a ticket must be sent by the buyer along with the instruction. The ticket has the form $E_K(T\|t)$, where T is the trace corresponding to the value instruction, t is a time stamp, $\|$ denotes concatenation and $E_K(\cdot)$ denotes encryption using a symmetric encryption algorithm with key K.*

3. [**Clearing**] *To clear a re-used value instruction, the merchant must send it to his bank along with a fresh ticket, i.e. a ticket that was not used in any previous clearing of the same value instruction. Freshness is ensured by the timestamp t in the ticket. The bank will not accept a re-used value instruction unless it is linked to a fresh ticket.*

The reader will notice that a re-used value instruction plus its ticket resemble a Millicent [7] piece of scrip and require a similar amount of processing. The advantage of our scheme over Millicent is that tickets are only required for re-used value instructions, which are typically a (small) fraction of the total number of instructions.

Note 8 (On clearing tickets). Clearing tickets one by one may cost too much as compared to the value of the associated value instruction. On the other hand, if the merchant chooses to clear tickets by batches, then he incurs in some risk of being paid with invalid tickets (*i.e.* not being paid at all). Note that tickets cannot be understood by the merchant (K is unknown to him), so the only way for the merchant to check the validity of a ticket is to try to have it cleared. Clearly there is a tradeoff between cost-effectiveness and security. It is up to the merchant's discretion to find the optimal strategy to solve this (economic) problem.

Note 9 (On coding backward branches). The structural coding used to encode instructions into traces requires that the targets of backward branches be signed (see Appendix). So each new iteration of a loop requires a run-time signature verification. This may be a computational burden if loops are too short, *i.e.* if they contain too few value instructions. Again, an economic tradeoff between verification costs and the flexibility offered by short loops is to be solved.

6   Conclusion

The new concept of spending program has been introduced in this paper. A spending program generalizes hash chains and hash trees used in several current micropayment schemes. The advantages of the new concept are twofold:

Flexibility Unlike hash chains, spending programs can offer multicurrency support, variable-valued coupons and selective use of such coupons. Case-like structures for choosing coupons at transaction-time can be implemented by the buyer or the buyer's smart card. One hot topic where this increased flexibility may be especially useful is intelligent autonomous payments by paying agents.

Efficiency With fixed-value coupons, the value of coupons must be the smallest fraction for the buyer to be able to pay any amount. This implies that verifying a lot of hashes is likely to be necessary for most micropayments. Variable-valued coupons permitted by spending programs allow micropayments of arbitrary value to be performed with a single hash verification. In fact, the distinction between micropayments and standard payments fades away with the availability of variable-valued coupons. On the other hand, iterative spending programs can be considerably more compact than hash chains and hash trees by allowing coupons to be re-used.

Regarding security, the structural coding mechanism sketched in the Appendix can be used to encode spending programs and obtain a degree of unforgeability similar to that of hash chains. However, we would like to stress that the concept of spending program is independent of the actual coding procedure used, as long as integrity is guaranteed.

References

1. R. Anderson, C. Manifavas and C. Sutherland, "NetCard - A practical electronic cash system", 1995. Available from author: Ross.Anderson@cl.cam.ac.uk

2. M. Bellare, J. Garay, R. Hauser, A. Herzberg, H. Krawczyk, M. Steiner, G. Tsudik and M. Waidner, "*i*KP - A family of secure electronic payments protocols", in *First USENIX Workshop on Electronic Commerce*, New York, July 1995.

3. CyberCash Inc., http://www.cybercash.com

4. J. Domingo-Ferrer, "Software run-time protection: a cryptographic issue", in *Advances in Cryptology - Eurocrypt'90*, ed. I. B. Damgaard, LNCS 473, Springer-Verlag, 1991, pp. 474-480.

5. J. Domingo-Ferrer, "Algorithm-sequenced access control", *Computers & Security*, vol. 10, no. 7, nov. 1991, pp. 639-652.

6. e-cash, http://www.digicash.com

7. S. Glassman, M. Manasse, M. Abadi, P. Gauthier and P. Sobalvarro, "The Millicent protocol for inexpensive electronic commerce", in *World Wide Web Journal, Fourth International World Wide Web Conference Proceedings*, O'Reilly, 1995, pp. 603-618.

8. R. Hauser, M. Steiner and M. Waidner, "Micro-payments based on *i*KP", IBM Research Report 2791, presented also at SECURICOM'96. http://www.zurich.ibm.com/Technology/Security/publications/1996/HSW96.ps.gz

9. C. Jutla and M. Yung, "PayTree: " Amortized-signature" for flexible micropayments", in *Second USENIX Workshop on Electronic Commerce*, Oakland CA, Nov. 1996.

10. J. Kohl, B. Neuman and T. Ts'o, "The evolution of the Kerberos authentication service", in *Distributed Open Systems*, eds. F. Brazier and D. Johansen, Los Alamitos, CA: IEEE Computer Society Press, 1994. See also *RFC 1510: The Kerberos Network Authentication System*, Internet Activities Board, 1993.

11. L. Lamport, "Password authentication with insecure communications", *Communications of the ACM*, vol. 24, no. 11, 1981, pp. 770-772.

12. T. P. Pedersen, "Electronic payments of small amounts", in *Security Protocols (April 1996)*, ed. M. Lomas, LNCS 1189, Springer-Verlag, 1997, pp. 59-68.

13. R. L. Rivest, *RFC 1321: The MD5 Message-Digest Algorithm*, Internet Activities Board, 1992.

14. R.L. Rivest and A. Shamir, "PayWord and MicroMint: Two simple micropayment schemes", Technical Report, MIT LCS, Nov. 1995.

15. R. L. Rivest, A. Shamir and L. Adleman, "A method for obtaining digital signatures and public-key cryptosystems", *Communications of the ACM*, vol. 21, no. 2, 1978, pp. 120-126.

16. Secure Electronic Transactions.
http://www.mastercard.com/set/set.htm

17. *Secure Hash Standard*, U. S. National Institute of Standards and Technology, FIPS PUB 180-1, April 1995.
http://csrc.ncsl.nist.gov/fips/fip180-1.txt

Appendix: Structural Coding for Program Integrity

We recall in this appendix the basic operating principles of the structural coding for program integrity given in [5]. Let $i_1, \cdots, i_n$ be a program, where i_k is a machine-language instruction (for a spending program, i_k can be either a value, a flow-control or an input-output instruction); i_n is not a branch (it can be set to be an **end** instruction). A normalized instruction format is defined: the length of all i_k is made equal by appending a known filler and then each i_k is appended a redundance pattern, whose length depends on the desired security level. Call the normalized program $I_1, \cdots, I_n$.

Let F be a one-way hash function, such that in general $F(X \oplus Y) \neq F(X) \oplus F(Y)$. Now, if $I_1, \cdots, I_n$ is a *sequential block* (containing no branches) it will be encoded into a *trace sequence* $T_0, T_1, \cdots, T_n$, where traces are computed in reverse order according to the following equalities:

$$T_0 = F(T_1) \oplus I_1$$
$$T_1 = F(T_2) \oplus I_2$$
$$\vdots \tag{3}$$
$$T_{n-1} = F(T_n) \oplus I_n$$
$$T_n = F(I_n)$$

Now consider branches. If I_k is a *forward unconditional branch* to I_j (*i.e.* $k < j$), it is translated as:

$$\cdots T_{k-1} = F(T_k) \oplus I_k$$
$$T_k = F(T_j) \oplus I_j$$
$$T_{k+1} = \cdots \tag{4}$$
$$\vdots$$
$$T_j = \cdots$$

When I_k is a *forward conditional branch* to instruction I_j, the following traces are computed (also in an index decreasing order):

$$\cdots T_{k-1} = F(T_k) \oplus I_k$$
$$T_k = F(T_{k'}) \oplus F(T_{k+1}) \oplus I_{k+1}$$
$$T_{k'} = F(T_j) \oplus I_j \tag{5}$$
$$T_{k+1} = \cdots$$
$$\vdots$$
$$T_j = \cdots$$

Backward branches from I_k to I_j (*i.e.* $k \geq j$) are slightly more complicated to encode, since the trace T_j corresponding to I_j cannot be used to compute T_k or $T_{k'}$ as in forward branches, because T_j follows T_k in the trace computation and depends on T_k (reverse trace computation). Let $\tilde{T}(j)$ a one-to-one integer function on j, for example the identity function. A *backward unconditional branch* at instruction I_k to instruction I_j is translated as:

$$\cdots T_{j-1} = F(\tilde{T}_j) \oplus F(T_j) \oplus I_j$$
$$\tilde{T}_j = F(T_j) \oplus \tilde{T}(j)$$
$$T_j = \cdots$$

$$\vdots \qquad\qquad\qquad (6)$$

$$T_{k-1} = F(T_k) \oplus I_k$$
$$T_k = F(\tilde{T}(j)) \oplus I_j$$
$$T_{k+1} = \cdots$$

Finally, the trace structure for a *backward conditional branch* is as follows:

$$\cdots T_{j-1} = F(\tilde{T}_j) \oplus F(T_j) \oplus I_j$$
$$\tilde{T}_j = F(T_j) \oplus \tilde{T}(j)$$
$$T_j = \cdots$$

$$\vdots \qquad\qquad\qquad (7)$$

$$T_{k-1} = F(T_k) \oplus I_k$$
$$T_k = F(T_{k'}) \oplus F(T_{k+1}) \oplus I_{k+1}$$
$$T_{k'} = F(\tilde{T}(j)) \oplus I_j$$
$$T_{k+1} = \cdots$$

In [5], the coding for non-recursive and recursive branch to subroutine is also detailed.

After the previous trace computation, it is up to the protection system (the buyer in the case of a spending program) to endorse the trace sequence with a signature on the first trace T_0 and on every $\tilde{T}_j$ in a backward branch target. So the protection system replaces T_0 with $t_0 = sk_{ps}(T_0)$ and the $\tilde{T}_j$'s with $\tilde{t}_j = sk_{ps}(\tilde{T}_j)$, respectively, where $sk_{ps}(\cdot)$ denotes encryption under the protection system's private key.

Now the trace sequence $t_0, T_1, \cdots, T_n$ is fed to a special processor instead of the executable instructions $i_1, \cdots, i_n$. This special processor contains a preprocessor for decoding the above trace structure and retrieving the executable instructions. The following properties hold:

Theorem 10 (Correctness). *The program $i_1, \cdots, i_n$ can be retrieved and executed from its corresponding trace sequence $t_0, T_1, \cdots, T_n$.*

The proof of Theorem 10 is a construction and it shows that, if the preprocessor is pipelined to the main processor, then there is no significant increase in the execution time. A few examples follow to give a flavour of the construction for decoding traces (details can be found in [4]):

- For a sequential instruction block, i_1 is retrieved by computing $I_1 = F(T_1) \oplus pk_{ps}(t_0)$ (where $pk_{ps}(\cdot)$ denotes encryption under the public key of the protection system); the rest of instructions i_k are retrieved by computing

$$I_k = F(T_k) \oplus T_{k-1}$$

  While i_k is being retrieved by the preprocessor, i_{k-1} can be executed by the main processor.
- If the instruction i_k is an unconditional forward branch to i_j, at cycle $k+2$, T_{k+2} is read, T_k is decoded by the preprocessor to obtain a valid instruction, and I_k is executed by the processor and recognized as an unconditional forward branch. At cycle $k+3$, only a read operation on T_j is performed; at cycle $k+4$ T_{j+1} is read, and T_k is evaluated by the preprocessor to obtain I_j

$$F(T_j) \oplus T_k = I_j$$

  Here $k+3$ and $k+4$ are idle cycles for the processor. From $k+5$ execution continues in sequence from I_j.
- For conditional forward branches, the decoding procedure depends on whether the condition is true or not. In the first case, it is very similar to decoding an unconditional forward branch. If the condition is false, decoding is very similar to the sequential instruction block case.
- For backward branches, decoding proceeds in a similar way, but a few more execution cycles are wasted.

Theorem 11 (Run-time integrity). *If a program $i_1, \cdots, i_n$ is stored as $t_0, T_1, \cdots, T_n$ and is decoded as described above, any instruction substitution, deletion or insertion will be detected at run-time, thus causing the main processor to stop execution before the substituted, deleted or inserted instruction(s). Moreover, only the last five read traces need be kept in the internal secure memory of the processor.*

The detailed proof of Theorem 11 can be found in [4]. The key point is that the one-wayness of the function F which connects traces makes it infeasible to successfully substitute, delete or insert traces ("successfully" means that no valid instructions can be decoded by the preprocessor beyond the point where traces have been substituted, deleted or inserted). It is important to stress that valid trace sequences can only be created by the protection system (the one who can sign with sk_{ps}).

Money Conservation via Atomicity in Fair Off-Line E-Cash

Shouhuai Xu[1], Moti Yung[2], Gendu Zhang[1], and Hong Zhu[1]

[1] Department of Computer Science, Fudan University, Shanghai, P. R. China.
{shxu, gdzhang, hzhu}@fudan.edu.cn
[2] CertCo, NY, NY, USA.
moti@cs.columbia.edu

Abstract. Atomicity and fault tolerance issues are important and typically open questions for implementing a complete payment scheme. The notion of "fair off-line e-cash" (FOLC) was originally suggested as a tool for crime prevention. This paper shows that FOLC schemes not just enable better control of e-cash when things go wrong due to "criminal suspicion" and other "regulatory/legal" issues, but it can also assure atomicity which takes care of conservation of money in case of failures during transaction runtime. The added protocols are very efficient and quite simple to implement. This kind of piggybacking atomicity control over "anonymity revocation" makes good sense as both actions are done by off-line invocation of the same trustees (**TTPs**). The resulting solution is a comprehensive yet efficient solution to money conservation in electronic cash transactions based on FOLC schemes. The adopted *recovery* approach makes the involved participants (customer, bank, merchant) sure that they can "re-think" the transactions when things go wrong, implying the atomicity of the transactions. We also take an optimistic approach achieving *fair exchange* costing only 2-round of communicational complexity (trivially the lower bound) with no additional **TTP** involvement since FOLC already employs such a party.

Keywords. Atomicity, Payments Systems, E-cash, Conservation of Money, E-Commerce Transactions Processing, Recoverability, Fair Off-Line E-Cash (FOLC)

1 Introduction

Any system should be robust enough to enable the completion of given tasks in the presence of faults and attacks. This is especially important in the context of e-cash based applications where certain messages actually represent real money. Task completion is not a major problem in face-to-face physical money transactions. In the context of account-based [ST95] and token-based [BGJY98] *on-line* e-commerce systems, where there is an *on-line* trusted third party (**TTP**), the issue is relatively easy to handle. On the other hand, in off-line e-cash based systems the issue becomes subtle. It is so because off-line e-cash schemes provide

M. Mambo, Y. Zheng (Eds.): ISW'99, LNCS 1729, pp. 14–31, 1999.

user anonymity, which implies that identifying parties via anonymity revoking for the sake of achieving *atomicity* or assuring *fair exchange* is totally unacceptable. This issue has never been dealt with in full generality and was only partially addressed by [CHTY96] (though with somewhat poor efficiency due to the use of generic methods which cause a heavily overloaded *on-line* **TTP** or log center) in the setting of (unconditionally anonymous) off-line e-cash based systems. The goal of the current paper is to present a much more efficient and complete solution to this problem. The approach we take is exploiting the properties of fair off-line e-cash (FOLC) schemes proposed in [FTY96, JY96].

We notice that the inefficiency of [CHTY96] is due to the adopted *two-phase commit* idea which is well-known in the area of database systems. While that tool is a powerful general purpose mechanism for atomicity in any transaction system, it is unnecessary in the context of FOLC based systems. As we will see, the efficiency of our solution comes from this observation: we need no "on-line" **TTP**, yet the functions implemented in [CHTY96] (say, fair exchange) are nevertheless provided. Our solution is comprehensive since we take the entire system into consideration. Moreover, we take an optimistic approach to *fair exchange* in the cost of only two-round communication (which is a trivial lower bound) while still without on-line **TTP**. Even in the occasional (and indispensable) invocation of the off-line **TTP** to handle a dispute, it can at times be casted by the party which is anyway responsible for such invocation due to an anonymity revocation request which is part of FOLC.

1.1 The Problem

Why hasn't off-line e-cash been adapted in practice so far? Many answers are possible, let us review some of the technical difficulties. Jakobsson and M'Raihi [JM98] have noticed that *conditional anonymity* and *card-ability* (whether they are efficient enough to be implemented using smart card) are two difficulties. Here we argue that *the assurance of conservation of money* (a kind of robustness or survivability) in e-cash transactions is another difficulty. Without promising money conservation, arguably nobody will dare to take the risk to use the systems as money can be lost and uncertainty prevails. This is justified by the analogous efforts for atomicity (and reduction of uncertainty about transaction completion) in generic distributed transactions in database systems.

Let us see some possible scenarios in off-line e-cash based systems. In the withdrawal protocol, what happens if the network crashes while the customer, **U**, is awaiting the e-coins? **B**, the bank, may have debited **U** [1] while **U** obtaining no coin. So, his [2] money is lost. On the other hand, **U** may have received the e-coins while claiming not to have received it! Thus, the money of the bank may be lost. We see that atomicity is needed to avoid both over-debiting and double-withdrawal.

---

[1] If **B** debits **U** after receiving the acknowledgment from **U**, then **U** may get coins without being debited (say, **U** will not send the acknowledgment). Even if special measures are taken, similar problems still exist.

[2] In this paper, customers are male, shop and **TTP** are female.

In a purchase transaction, both the customer and the shop may agree on the price and electronic goods [3] after some negotiation. Seeing the encrypted e-goods from the shop, the customer optimistically sends his e-coins to her and waits for (her share of) the decryption key (see section 4 and 5). What is the dilemma for them if the network crashes (or when the customer's PC malfunctions, or even when there is a denial-of-service attacker) while neither U is acknowledged by the shop (S) for his e-coins nor S is acknowledged by U for her share of the decryption key [4]? Due to the complexity of the causes, there are two possible states for this transaction.

- S has received the e-coins. Since U does not know the state of the transaction, what can U and S do? (1) S will deposit the e-coins. If U spends these coins at another time, he is bound to face an accusation of double-spending in the scheme of [FTY96] (alternatively, possibly over-spending in the scheme of [JY96][5]). So, U better dare not spend the coins and, effectively, his money is lost. (2) S will not deposit the e-coins as she has received no acknowledgment for the key from U, effectively she sells her e-goods without income (a denial-of-service attacker may gain benefit from it as well). As a result, her money is lost.
- S has not yet received the e-coins so she trivially will not deposit them. However, U still dare not spend his e-coins at another transaction because he fears taking the risk of being accused a double-spender or an over-spender.

Furthermore, even if U knows that S will not deposit the coins if she has not received the acknowledgment for the key, though now he will not face an accusation of double-spending (or over-spending) when he spends the e-coins again, his ID may still be revealed in some schemes (say, [FTY96]) by colluding shops.

While, intuitively, the broken transaction can be recovered through a special protocol, how about if the transaction information (including the URL, price, contract) is lost due to either hardware or software crashes of the customer's poor PC? He can not recover it even if he wants to.

In the deposit, the shop should be assured that when she sends the e-coins, she is bound to be credited correspondingly. What happens if the network crashes

---

[3] In this paper we focus on e-goods while the ideas can also be adapted to the setting of physical goods.

[4] We note that the reliability mechanism at the transport layer (say, TCP) will not help us in this case. The reason being that the verification for the validity of either the e-coins or the decryption key (therefore the decrypted e-goods) is executed at the application layer, so the acknowledgment must also be from the peer application layer entity. Indeed, the acknowledgment for the valid e-coins is always the transmission of (the share of) the decryption key itself in practice. On the other hand, the underlying connection itself may have been released due to a timeout.

[5] For example, if we assume that the avaliable value of a coin is \$4 while its denomination is \$10 and the failure transaction is of \$2, if he spends at another transaction \$3, then he may be accused of being an over-spender and his anonymity is revoked.

before **B** acknowledging **S**? **S** will not be sure if, and how much, her account has already been credited correctly.

To summarize, numerous mishaps and problems occur when atomicity of a payment related transaction is not assured.

1.2 Our Solution

In this paper we present an efficient solution to the above problem via *withdrawal atomicity, payment-delivery atomicity* [6], and *deposit atomicity* in FOLC schemes [FTY96, JY96] based transactions. We assure that while the security and anonymity of the original schemes are still maintained, our solution is efficient in that there needs no (additional) on-line **TTP** even in the dispute handling protocol [7] for fair exchange.

This paper, therefore, shows that FOLC schemes do not just enable better control of e-cash when things go wrong due to "criminal suspicion" and other "regulatory/legal" issues. Rather, the same parties that take care of this can also assure atomicity and conservation of money in case of failure during the transactions runtime. This kind of piggybacking atomicity control over "anonymity revocation" makes good sense as both actions are done by off-line invocation of the same **TTP**s. This implies that FOLC schemes ease atomicity in e-cash systems.

It has also been noted in [W98] that the similarity between database transactions and commercial ones, unfortunately, always misleads people to simple minded implementations. While the traditional destructive *prevention-and-recovery* approach (say, two-phase commit [BN97]) to *atomicity* in database community is powerful enough, it provides poor efficiency in our e-cash context (as we mentioned above). Instead, our simple *recovery* philosophy makes it much more efficient than the traditional approach. We comment that just as mentioned in [CHTY96], the basic properties of transaction systems which accompany atomicity: CID (which stands for: Consistent, Isolated, and Durable, refer to [BN97] for details), are naturally satisfied in e-cash transactions and therefore are omitted in the rest of this paper.

This paper also shows that the well known problem of *fair exchange* (refer to [ASW98]) is not so hard to implement in the context of "exchange fair off-line e-coin for e-goods". In our solution, the *fair exchange* need only two rounds of communication in real-life. That is, one sends the money, and the other sends the e-goods (or the share of the decryption key). It is even better than the 3-round protocol most recently proposed by Bao et al. in [BDM98]. Furthermore, 2-round

---

[6] As we focus on a total solution to e-goods based e-commerce, we incorporate delivery into payment. In the context of physical goods, the payment atomicity is naturally implied by the solution of this paper. As for fair exchange or atomic delivery, it is obviously out of the reach of the e-system by itself.

[7] On one hand, according to [BP89], pure fair exchange is unachievable and dispute handling must come to rescue in electronic commerce. On the other hand, the off-line **TTP** in this paper can be performed by any Trustee for conditional anonymity in the original schemes.

communicational complexity is the trivial lower bound for any fair exchange protocol as, even in the real world with physical cash, it is still necessary. The trick is that the (off-line) bank plays a key role, i.e., certifying the reciept of money.

1.3   Outline

In section 2 we briefly review the related work. The basic ideas, definitions, and assumptions used in this paper are presented in section 3. Our atomicity providing extensions to the electronic coin schemes of [FTY96, JY96] are proposed respectively in sections 4 and 5. We conclude in section 6.

2   Related Work

2.1   E-Cash Schemes

The basic electronic cash system consists of three (probabilistic polynomial-time) parties: bank **B**, user or customer **U**, and shop or merchant **S** , and three main procedures: withdrawal, payment, and deposit. Users and shops maintain their respective accounts with the bank, while:

- **U** withdraws electronic coins (digital cash) from his account, by performing a withdrawal protocol with the bank **B** over an authenticated channel.
- **U** spends a coin by participating in a payment protocol with a shop **S** over an anonymous channel.
- **S** performs a deposit protocol with the bank **B**, to deposit the user's coin into her account.

The original electronic cash aimed at offering some level of user anonymity during a purchase with the intention to emulate electronically the properties of physical cash exchange. The first generation of electronic cash systems are proposed in e.g., [CFN88, OO91, FY93, B93, F93]. These schemes provide anonymity in information theoretic sense through the useful primitive called blind signature [C82] or its variants. Realizing that the complete anonymity can be abused to commit "perfect crimes" (money laundering, blackmailing, bribery taking, even bank robbery [vSN92, JY96]) a way to avoid perfect anonymity was sought out. In [BGK95, SPC95, CPS96, JY96, J97, FTY96, T97, CFT98], conditional anonymity (alternatively, fair off-line digital cash, revokable anonymity) schemes with various extensions have also been proposed. It seems that while the first generation of e-coin schemes focuses on privacy, the second one pays much attention to balance privacy and robustness (against abuse), and the notion of trustees who may revoke anonymity was put forth.

The *indirect discourse proofs* technique was used by Frankel, Tsiounis, and Yung in [FTY96] to offer the Bank the confidence that the Trustee will be able to trace in case tracing is needed. Further such schemes are in [DFTY97,dST98,

FTY98]. Another recent extension is presented in [T97, CFT98] where the divisibility properties from [OO89, OO91, O95] are incorporated. At the same time, Jakobsson and Yung presented a versatile electronic cash scheme in [JY96], and its *magic ink signature* version in [JY97, J97]. This latest scheme tries to cope even with the strong bank robbery attack by introducing trustees outside the bank system.

2.2 Previous Approaches to Atomic E-Commerce Transactions

Tygar [T96] is the one who first noticed the problem of *atomicity*, then [CHTY96] is the first work aimed at supporting atomicity for electronic commerce transactions based on off-line e-cash schemes. Though the authors used a heavily overloaded inefficient on-line **TTP** (called log center), this is nevertheless the first solution realizing the three classes[8] of atomicity proposed in [T96], namely:

- **Money atomicity**: Transactions feature atomic transfer of electronic money, i.e., the transfer either completes entirely or not at all;
- **Goods atomicity**: Transactions are money atomic and also ensure that the customer will receive goods iff the merchant is paid;
- **Certified delivery**: Protocols are goods atomic and also allow both the customer and merchant to prove exactly what was delivered. If there is a dispute, this evidence can be shown to a judge to prove exactly what goods were delivered.

This problem has been paid attention to also in the easier settings of online schemes. For example, atomicity for on-line account-based transactions is addressed in [ST95], whereas atomicity for on-line token-based transactions is given in [BGJY98].

Following are a few remarks regarding the early work:

1. The solution proposed in [CHTY96] is weakened by the fact that they use an impractical variant of the blind coin from [C82, CFN88] (first generation e-cash) as their building block. It is well known that these schemes are subject to the potential trouble of money laundering, blackmailing and bribery committing attack (to overcome which FOLC was invented).

2. Furthermore, the solution proposed in [CHTY96] is not double-withdrawal robust. For example, the customer claims after receiving the e-coins that he has never seen them. Due to the perfect anonymity of the underlying blind signature, the bank is unable to trace the e-coins. Therefore, the money of the bank is lost.

3. It is already mentioned in[CHTY96] that the "ripping coin" idea proposed by Jakobsson in [J95] is not *money atomic*. This is because each of the half protocols themselves may be interrupted, leaving the digital cash again in an ambiguous state (alternatively, in a blocking state, using the terms of the area of transaction processing [BN97]).

4. The ideas proposed in [BBC94] against lost of money due to wallet loss or malfunctioning can be used as an orthogonal building block, and be added to

---

[8] The last one is partially implemented.

our solution. That is, we concentrate on the protocol transactions and we do not consider the possibility of money lost resulted from crashes of the e-coin wallets. We assume this kind of fault tolerance is provided.

3   Preliminary

3.1   Basic Ideas

As mentioned above, the users of an electronic commerce system must be assured (in addition to security and privacy) that each transaction executes either completely or not at all (i.e., *atomicity*) implying that their money will not be lost due to the system itself. As any electronic commerce application has to involve some delivery process, we incorporate delivery of e-goods into payment in this paper resulting in an efficient payment-delivery scheme. We attempt to realize a comprehensive atomicity (including *withdrawal atomicity*, *payment-delivery atomicity*, and *deposit atomicity*) as non-atomicity may occur at any phase due to any malfunctioning within the system or even an external (denial of service) attack. The basic philosophy used in our solution is *recovery* rather then *prevention-and-recovery* of the database community (as in [CHTY96]) for the sake of efficiency.

While we use a simple "resolve protocol" for *payment-delivery atomicity*, a "dispute handling protocol" for fair exchange in any protocol (even in the real world [BP89]) is also adopted. We still present a secure (against denial-of-service attack) "refund function" in case the customer needs to either refund his e-coins or exchange them for fresh e-coins in certain e-cash schemes.

Definition 3.1 *Let τ_0, τ be global time, Δ is some (implicitly or explicitly) pre-designed period. Let us assume that an off-line e-coin based transaction begins at τ_0, whereas $\tau = \tau_0 + \Delta$. A withdrawal protocol is said to have withdrawal atomicity if the customer's account is debited iff he receives the corresponding e-coins within τ. A payment-delivery protocol is said to have payment-delivery atomicity if the payee receives the coins iff the payer receives the e-goods (namely, the decryption key) within τ. A deposit protocol is said to have deposit atomicity if the payee is credited with certain amount as well as the bank receives the corresponding e-coins within τ.*

Remarks:
1. We focus on the "practical value" of a coin rather than its "face value". Namely, the customer in a withdrawal protocol may get some additionally legal coins (in a mathematical sense), but he dare not spend them as he is bound to face the *tracing of owner* (refer to **Theorem 2** and **4**). We note again that, as mentioned before, the withdrawal protocol in [CHTY96] is not double-withdrawal robust. Another case, in a payment-delivery protocol, the shop may "obtain" some coins and will not deposit them. It is said that she does not receive the coins as the "practical value" of the coins is zero though in a mathematical

sense they are valid. It is interesting to note that its mirror exists in the setting of physical cash in the real world.

2. Obviously, if we consider only a payment protocol, the corresponding definition for atomicity can be easily adapted from the above one for payment-delivery. This is omitted in the rest of this paper due to space limitation.

3. The *money atomicity* defined in [T96] is trivially implied by our definitions. (Thus, we will not explicitly discuss it.) Consequently, the assurance for the conservation of money and fair exchange (i.e., *goods atomicity*) in e-cash based transactions is indirectly implied by our definitions. In the context of this paper, the so-called *certified delivery* is directly implemented by the transaction atomicity and fair exchange in a "destructive" way which means that, though the customer can abuse the complaint facility, he can obtain no additional useful advantage (e.g., e-goods) whereas the shop has to be honest since the bank can present the records of her coin income. This suggests that carefully designed protocols can prevent (rather than resolve after the fact) some abuses in protocols for certain complicated problems (e.g., certified delivery) at least in certain settings (say, e-coin vs. e-goods).

3.2   Assumptions

All the computational complexity assumptions for security of the transactions are naturally inherited from the underlying e-cash schemes [FTY96, JY96], therefore, they are claimed respectively in the context while necessary or omitted while it is clear from the context.

Additionally, we assume that there exist anonymous communication channels for anonymity or privacy. Such a channel is typical in the literature, say, [CHTY96, FTY96, T97, CFT98, JY96, J97, JM98][9].

We assume that a transaction may be broken in an arbitrary fashion. It is also assumed that the money of the customer will not be lost due to the crash of his PC, as he has a fault tolerant e-wallet of [BBC94]. In other words, while the transaction information may be lost due to the malfunctioning of his PC, his money is still available. On the other hand, the shop server is assumed to be powerful enough to address either hardware or software malfunctioning. This is plausible due to the fact that many fault tolerant mechanisms for servers are available.

4   Atomicity Extension to [FTY96, T97, CFT98] E-Cash Scheme

4.1   The Extended Payment-Delivery Protocol

We first present a small extension of the original payment protocol to incorporate the delivery mechanism. It is easy to see that this extension compromises neither

---

[9] It should be noted that, no anonymous channel is needed in the context of the original [JM98]. However, here we consider the whole process in an electronic commerce application rather than solely the payment protocol as in [JM98].

security nor privacy. The original protocols are copied in **Appendix 1**. Though the original deposit protocol also needs to be updated to include checking for the validity of the payment (not yet expired), due to space limitation, we omit this extension here.

1. **U** and **S** negotiate through some interaction, and agree on the price, the goods description *desc*, and invalid date/time $\tau = \tau_0 + \Delta$ where τ_0 is the time that the interested transaction begins and Δ is the negotiated lifetime for it. Alternatively, τ is the deadline for **S** to deposit the received e-coins.
2. **U** sends **S** his temporary public key with intention to generate a standard DH key k [DH76] for the encryption (and decryption) of the e-goods[10].
3. **S** sends the encrypted goods $E_k(goods)$ to **U**, where E_k is a public, secure against chosen message attack, symmetric cipher algorithm.
4. **U** sends to **S**: $A_1 = g_1^{u_1 s}, A_2 = g_2^s, A, B, (z, a, b, r)$.
5. **S** checks $A = A_1 A_2$, $A \neq 1$, $sig(A, B) = (z, a, b, r)$, and then responses to **U**: $d = H_1(A_1, B_1, A_2, B_2, I_S, date/time, desc, \tau)$ (where I_S is the shop's ID).
6. **U** computes $r_1 = d(u_1 s) + x_1$, $r_2 = ds + x_2$, and then sends r_1, r_2 to **S**.
7. **S** accepts iff $g_1^{r_1} = A_1^d B_1$ and $g_2^{r_2} = A_2^d B_2$, and then sends her share of the DH key k (i.e., her temporary public key) to **U**.

4.2   The Solution

The intuitive solution, as discussed earlier, is to activate some recovery protocol in which the participants re-think the transcript. While this is natural, indeed, for withdrawal and deposit protocols, it is subtle for the payment-delivery protocol (see below). On the other hand, double-withdrawal mentioned above can be blocked using the original *coin tracing* technique in [FTY96, T97] (which is another reason to build upon FOLC). The rationality behind this choice of recovery at which phases comes partially from the fact that in withdrawal **U** has to authenticate himself to establish the ownership of his account, and in deposit the shop's account is embedded in the transcript, whereas the state of a non-atomic purchase transaction (in general) may be very complicated.

As argued in subsection 1.1, **U** and **S** will face a dilemma if the network crashes (alternatively, the customer's PC malfunctions, or there is a denial-of-service attacker) while neither **U** is acknowledged by **S** for his e-coins nor **S** is acknowledged by **U** for her share of the decryption key. Regardless of whether **S** has received the e-coins or not, if **U** does not lose the state information for the transaction, he can activate simply a recovery transaction. Otherwise, he needs to activate a resolve protocol with **TTP** and **B** to see if his coins has ever been deposited within the negotiated τ. If so, **S** is asked to send **U** the goods

---

[10] k is chosen by only one party, either **U** or **S**, the system is subject to denial-of-service attack if the corresponding recovery protocol is not carefully designed (i.e., without requesting **U** to prove the ownership of the e-coins concerned). However, in this paper the notations "share of the decryption key" and "decryption key" are interchangeable.

(otherwise she will face a legal action); if not, **U** has to refund the e-coins for the sake of anonymity.

However, there are still two problems yet to be addressed. One is that when a claimed customer recovers a purchase transaction, if he wants to change the original share of the negotiated DH key k used for the encryption (therefore decryption) of the e-goods (say, he may claim that he has lost his original share of k), a protocol for "ownership proof" of the e-coins has to be activated. Otherwise, a denial-of-service attacker can snatch the e-goods. Another, is the security of the refund protocol in which the "ownership proof" for the e-coins is also necessary to frustrate the determined denial-of-service adversary. To block the two possible attacks, for the sake of efficiency, a proof for knowledge of Schnorr scheme [S91] can be initiated. For example, a Schnorr proof for knowledge $\log_{g_1} B_1 = x_1$ is competent. In the rest of this paper, this "ownership proof" is used as a primitive without further detailed.

Recovery protocol for *withdrawal atomicity*:

1. **U** proves to **B** that he is the owner of certain account.
2. **U** sends to **B** the transcript a', b', c' of the broken transaction (alternatively, if **U** wants to generate new coins as he claimed that the state information has been lost and then he is unable to present it, or **U** claims that he has received no e-coins, or **B**'s signature key has just been replaced, the original coins should be blacklisted using the original *coin tracing* technique of [FTY96] to realize double-withdrawal robustness).
3. Otherwise, after checking that there really exists this session, **B** responses with $r' = c'x + w$ to **U**. If it has not yet debited, it does so now.

A replaced signature key of the bank should be still valid for a period of time in which coins signed by it are acceptable. Within this period the recovery protocol above can also be used to "refresh coins" of users, while merchants are assumed to rush to deposit such coins before they become invalid.

If a payment-delivery protocol is broken, **U** has two choices: either to recover or to resolve it through the following Recovery and Resolve protocols respectively.

Recovery protocol for *payment-delivery atomicity:*

1. **U** sends the same transcript in the broken payment-delivery protocol to **S** (alternatively, if he wants to change the original share of k, the protocol for the proof of ownership of the coins is initiated).
2. After checking the validity, **S** sends her share of k to **U**.

Recovery protocol for *deposit atomicity*:

1. **S** sends **B** the transcript of the not yet acknowledged deposit.
2. **B** checks the existence and the state (i.e., whether or not has been cancelled due to expiration) of the claimed session, and does whatever as this is a normal run of deposit protocol. All coins not yet credited are credited now.

If **U** does not get the share of k from **S** until the invalid date/time τ (no matter he has initiated the Recovery protocol or not), he can now activate the resolve protocol.

Resolve protocol: (performed between **U**, **S**, **B** and **TTP** over an anonymous channel, the **TTP** can be any one chosen from the set of trustees responsible for revocation of anonymity)

1. **U** sends **TTP** the transcript of the broken payment-delivery transaction (if he has lost the state information of the transaction, he can simply present the coins).
2. After checking that the payment has expired (otherwise, **S** may still honestly manage to send her share of k to **U** before τ), **TTP** first asks **U** to prove his ownership of the e-coins using the above mentioned "ownership proof" primitive (as **S** may need to refund). **TTP** asks **B** to see if these coins have ever been deposited. If yes, **TTP** asks **S** to send her share of k to **U** otherwise she will face a legal action; if not, this payment is cancelled and the coins are refunded. (Alternatively, **U** and **B** can activate another instance of withdrawal protocol to exchange these coins for fresh coins blindly.)

Remarks: 1. In withdrawal and deposit protocols we use implicit time while we use explicit one in payment-delivery protocol. The rationality is that sometimes delivery may not be completed immediately. However, it is necessary for **S** to deposit all received coins before their invalid date/time τ.

2. It is interesting to notice that the original payment protocol in [B93, FTY96] used the information *date/time* (it is the τ_0 for our purpose) however, it is not intended to address the problem of atomicity. The synchronous clock used in [B93, FTY96] can also be replaced with the global standard time.

3. The I_S used in the payment is important as there is no measure for **S** to prove that she is the coin's owner.

4. The shop's share should be consistent with the k of $E_k(goods)$, otherwise the customer will not be able to correctly decrypt it. (If she intends to deny the contents/semantics of the goods, *desc* will frustrate her as it may include a cryptographic hash of the content!). Though **S** changing her share will not result in any security or privacy vulnerability, this may need to be avoided unless it is compromised (say, due to a Trojan Horse) because the goods may be large and resuming the transmission from the broken point may become necessary (just like the function supported by most of the current FTP tools). Notice also that **S**'s temporary public key used in different transactions are different (so the public key may be used to encrypt an actual content key, in case we do not want to change the content's encryption).

4.3   Claims

We show that, while our solution inherits all the properties proved in [FTY96] (therefore, the extensions in [T97, CFT98]), it also is atomic (therefore, fair exchange and conservation of money are assured).

Theorem 1 *The extended fair off-line e-cash scheme satisfies unreusability, unforgeability, unexpandability and (conditional) untraceability.*

The proof is almost the same as [T97]. The rationality is that in random oracle model, the distribution

$$H_1(A_1, B_1, A_2, B_2, I_S, date/time, desc, \tau)$$

used in the payment is identical to the original

$$H_1(A_1, B_1, A_2, B_2, I_S, date/time).$$

Theorem 2 *The above extended scheme based transactions are of withdrawal atomicity, payment-delivery atomicity, and deposit atomicity. The assurance for the conservation of the money is implemented by them (and the recovery and resolve protocols if necessary).*

We only give the proof for *payment-delivery atomicity*, since the rest are quite trivial.

proof: (sketch) When we consider a transaction in which there are two receivers (one for money and the other for e-goods), there are only four cases:

1. Everyone succeeds in receiving the intended items. It means that there is a successful transaction and, of course, *payment-delivery atomicity*.
2. **S** received (i.e., having already been credited to her account[11]) the money and **U** did not receive the share of k from **S**. In this case, **U** can activate the resolve protocol. As long as **S** has deposited the coins, **S** has to send her share of k to **U**, otherwise **S** will face a legal action. (Obviously, the goods description *desc* is useful when **U** has lost all the state information.)
3. **U** received the goods and **S** did not receive the coins. This is impossible since we assume that **S** sends her share of k after seeing the coins and she has a powerful fault tolerant system for reliability.
4. Both **U** and **S** receive nothing. In this case the resolve protocol will assure the customer that his money is conserved.[12]

□

It is easy to see that privacy for the customer is preserved throughout the processes.

It is interesting to note that the well known problem of fair exchange with off-line **TTP** (refer to [ASW98]) is not as difficult as may have been imagined in the context of "fair exchange e-coins for e-goods" of the current paper. The rationality behind this is that in this context the bank **B** plays an important role.

---

[11] It is plausible to assume that **S** is bound to deposit the coins she has received.

[12] **S** may not deposit the e-coins as she has not yet sent her share of the decryption key to **U**. In this case, she has e-coins only in a mathematical sense within the deadline τ.

5 Atomicity Extension to [JY96, J97] E-Cash Scheme

5.1 The Extended Payment-Delivery Protocol

Due to the limitation of space, we only focus here on the original payment protocol. The problems arising at withdrawal and deposit can be addressed using the ideas similar to the ones in the last section. The difference is that the protocol used to prove the ownership of the coins may be different. A practical proof can be constructed from the concrete implementation of the original cryptographic primitive (see **Appendix 2**) in which (x, y) is the corresponding pair of secret and public keys. A possible solution is a challenge in which value 0 and a random number are included. It should also be noted that this proof is necessary only when the customer wants to change his share of the original key k or refund a coin due to reasons resulted from non-atomicity (though refunding is unnecessary in this system, the customer may really want to do it).

1. **U** and **S** negotiate and agree on the price, the goods description $desc$, and invaliddate/time $\tau = \tau_0 + \Delta$ where τ_0 is the time that the interested transaction begins and Δ is the negotiated lifetime for it. Alternatively, τ is the deadline for **S** to deposit the received e-coins.
2. **S** sends the encrypted goods $E_k(goods)$ to **U**, where k is a DH key as mentioned above, E_k is a public, secure against chosen message attack, symmetric cipher algorithm.
3. **U** sends to **S**: (y, s).
4. **S** checks $s = S_B(S_O(y))$ and sends back challenge c (in which $desc$ and τ are included) [13] to **U**.
5. **U** sends the answer $a = S_x(c)$ to **S**.
6. **S** checks $V_y(a, c) = 1$ and keeps the transcript (y, s, c, a) and then her share of k is sent to **U**.

5.2 The Solution

Here we only give the recovery protocol for payment-delivery and the resolve protocol since they are very different from the ones in the last section.

   Recovery protocol for *payment-delivery atomicity*:

1. **U** sends the transcript in the broken payment-delivery protocol to **S** (alternatively, if he wants to change his original share of k, the protocol for "ownership proof" of the coins is initiated).
2. After checking the validity of the transcript as in the last section, her share of k is sent back to **U**.

   Similarly, **U** can activate the following resolve protocol if necessary.

---

[13] the challenge sent by **S** is composed of some part of a predetermined form (challenge semantics) and a random string.

Resolve protocol:

1. U sends **TTP** the transcript in payment-delivery transaction (if he has lost the state information of the transaction, he may simply present the coins).
2. After checking that the payment has expired, **TTP** first asks **U** to prove his ownership for the e-coins using the abovementioned primitive "ownership proof" in case he wants to refund. Then **TTP** asks **B** to see if those coins have ever been deposited. If yes, **TTP** asks **S** to send her share of k to **U** otherwise she will face a legal action; if not, this payment is naturally cancelled and **U** may not need to refund.

Remarks: 1. In [JY96], the authors presented a "user complaints" protocol. However, it is used for the user to complain that the coins received are not correctly constructed.

2. We recommend that it is better to incorporate the identity of **S** into the challenge as she can not prove the ownership of the coins while deposit. In this case, even the communication channel in deposit protocol is corrupted, the bad guy is still unable to get the coins via denial-of-service attack.

5.3   Claims

While our solution inherits all the properties proved in [JY96, J97], it is also atomic as we show below (implying, fair exchange and conservation of money). From [JY96, J97] and the fact that the basic protocols do not change, we get:

Theorem 3 *The extended electronic cash scheme satisfies unforgeability, impersonation safety, overspending detection, overspending robustness, traceability, revocability, anonymity, framing-freeness, and refundability.*

Theorem 4 *The above extended scheme based transactions are of withdrawal atomicity, payment-delivery atomicity, and deposit atomicity. The assurance for the conservation of the money is implemented by them (and the recovery and resolve protocols if necessary).*

Again, we only give the proof for *payment-delivery atomicity*, since the rest are easy.

proof: (sketch) When we consider a transaction in which there are two receivers (one for money and the other for e-goods), there are only four cases:

1. Everyone succeeds in receiving the intended items. It means that there is a successful transaction and, of course, *payment-delivery atomicity*.
2. S received (i.e., having already been credited to her account) the money and U did not receive her share of k from S. In this case, U can activate the resolve protocol. As long as S has deposited the coins, S has to send her share of k to U, otherwise S will face a legal action. (Obviously, the goods description *desc* is useful when U has lost all the state information.)

3. **U** received the goods and **S** did not receive the coins. This is impossible as we assume that **S** sends her share of k after seeing the coins and she has a powerful fault tolerant system for reliability.
4. Both **U** and **S** receive nothing. In this case the resolve protocol will assure the customer that his money is conserved.

□

6 Conclusion

We have presented a comprehensive, secure, anonymous, yet efficient solution to conserve the amount of money via atomicity in fair off-line e-cash based e-commerce transactions. The adopted *recovery* approach employs simple protocols and at the same time it assures the involved participants that they can re-think the transactions when things go wrong, thus the atomicity of the transactions. We also take an optimistic approach to *fair exchange* which can be done in only 2-round of communication (which is trivially the lower bound). All of this is done with no additional **TTP** involvement since there have already been ones for the possibly requested revocation of anonymity in the system. Consequently, our solution assures conservation of money and fair exchange through *withdrawal atomicity, payment-delivery atomicity*, and *deposit atomicity* whereas the security and anonymity of the original schemes are naturally inherited.

This paper shows the strength of the setting of fair off-line e-cash schemes (FOLC). It not just enables better control of e-cash when things go wrong due to "criminal suspicion" and other "regulatory/legal" issues, but the same parties that take care of this can also assure atomicity and control of conservation of money in case of failure during the transactions. This kind of piggybacking atomicity control over "anonymity revocation" fits nicely since both actions are done by off-line invocation of the same **TTP**s. The obvious conclusion is that fair off-line e-cash schemes ease atomicity in e-cash systems.

Acknowledgments: Many thanks to Markus Jakobsson for very valuable discussions. We also thank the anonymous referees for useful comment. The first author also thanks Feng Bao, Jan Camenisch, David Chaum, Ronald Cramer, and Wenbo Mao for their valuable feedback.

References

[ASW98]   N. Asokan, V. Shoup, and M. Waidner, Optimistic Fair Exchange of Digital Signature, Eurocrypt'98
[B93]   S. Brands, Untraceable Off-Line Cash in Wallets with Observers, Crypto'93
[BBC94]   J. Boly, A. Bosselaers, R. Cramer et al., The ESPRIT Project CAFE: High Security Digital Payment Systems, ESORICS'94

[BDM98]   F. Bao, R. Deng, and W. Mao, Efficient and Practical Fair Exchange Protocols with Off-Line TTP, IEEE Security and Privacy, 1998

[BGJY98] M. Bellare, J. Garay, C. Jutla, and M. Yung, *VarietyCash*: A Multi-Purpose Electronic Payment System (Extended Abstract), Usenix Workshop on Electronic Commerce'98

[BGK95]   E. F. Brickell, P. Gemmell, and D. Kravitz, Trustee-Based Tracing Extensions to Anonymous Cash and the Making of Anonymous Change, SODA'95

[BN97]   P. A. Bernstein and E. Newcomer, Principles of Transaction Processing, Morgan Kaufmann Publishers, Inc., 1997

[BP89]   H. Burk and A. Pfitzmann, Digital Payment Systems Enabling Security and Unobserability, Computer & Security, 8/5, 1989, 399-416

[C82]   D. Chaum, Blind Signatures for Untraceable Payments, Crypto'82

[CFN88]   D. Chaum, A. Fiat, and M. Naor, Untraceable Electronic Cash, Crypto'88

[CFT98]   A. Chan, Y. Frankel, and Y. Tsiounis, Easy Come-Easy Go Divisible Cash, Eurocrypt'98

[CHTY96] J. Camp, M. Harkavy, J. D. Tygar, and B. Yee, Anonymous Atomic Transactions, 2nd Usenix on Electronic Commerce, 1996

[CPS96]   J. Camenisch, J. Piveteau, and M. Stadler, An Efficient Fair Payment System, ACM CCS'96

[DFTY97] G. Davida, Y. Frankel, Y. Tsiounis, and M. Yung, Anonymity Control in e-cash. In the 1-st Financial Cryptography, LNCS 1318 Springer.

[dST98]   A. de Solages and J. Traore, An Efficient Fair off-line electronic cash with extensions to checks and wallets with observers, In the 2-d Financial Cryptography.

[DH76]   W. Diffie and M. E. Hellman, New Directions in Cryptography, IEEE Transactions on Information Theory, 1976, 644-654

[F93]   N. Ferguson, Extensions of Single-Term Coins, Crypto'93

[FTY96]   Y. Frankel, Y. Tsiounis, and M. Yung, Indirect Discourse Proofs: Achieving Fair Off-Line E-Cash, Asiacrypt'96.

[FTY98]   Y. Frankel, Y. Tsiounis, and M. Yung, Fair Off-line e-cash made Easy, Asiacrypt'98.

[FY93]   M.K. Franklin and M. Yung, Secure and Efficient Off-line Digital Money, ICALP'93 LNCS 700, Springer Verlag. 1993.

[J95]   M. Jakobsson, Ripping Coins for a Fair Exchange, Eurocrypt'95

[J97]   M. Jakobsson, Privacy vs. Authenticity, PhD thesis, 1997

[JM98]   M. Jakobsson and D. M'Raihi, Mix-based Electronic Payments, Workshop on Selected Areas in Cryptography, 1998

[JY96]   M. Jakobsson and M. Yung, Revokable and Versatile E-Money, 3rd ACM Computer and Communication Security, 1996

[JY97]   M. Jakobsson and M. Yung, Magic Ink Signature, Eurocrypt'97.

[O95]   T. Okamoto, An Efficient Divisible Electronic Cash Scheme, Crypto'95

[OO89]   T. Okamoto and K. Ohta, Disposable Zero-Knowledge Authentication and Their Applications to Untraceable Electronic Cash, Crypto'89

[OO91]   T. Okamoto and K. Ohta, Universal Electronic Cash, Crypto'91

[S91]   C. P. Schnorr, Efficient Signature Generation by Smart Cards, J. Cryptology, 4(3), 1991, 161-174

[SPC95]   M. Stadler, J. M. Piveteau, and J. Canmenisch, Fair Blind Signature, Eurocrypt'95

[ST95]   M. Sirbu and J. D. Tygar, NetBill: An Internet Commerce System, IEEE Compcon'95

[T96]   J. D. Tygar, Atomicity in Electronic Commerce, ACM Symposium on·Principles of Distributed Computing, 1996

[T97]   Y. S. Tsiounis, Efficient Electronic Cash: New Notions and Techniques, PhD thesis, 1997

[vSN92]   B. von Solms and D. Naccache, On Blind Signatures and Perfect Crimes, Computers & Security, 11(6), 1992, 581-583

[W98]   M. Waidner, Open Issues in Secure Electronic Commerce, 1998

Appendix 1   The Protocols of [FTY96]

All the notations used here are the same as in [FTY96]. Briefly, G_q is a prime order q subgroup of Z_p where p is also a prime. g, g_1 and g_2 are generators of G_q. H, H_0, H_1, ..., are hash functions of collision intractable. $\mathbf{B}$ chooses her secret $x \in_R Z_q$. $\mathbf{B}$ publishes $p, q, g, g_1, g_2, (H, H_0, H_1, ...,)$ and her public keys $h = g^x$, $h_1 = g_1^x$, $h_2 = g_2^x$.

When $\mathbf{U}$ setups an account with $\mathbf{B}$, $\mathbf{B}$ associates $\mathbf{U}$ with $I = g_1^{u_1}$ where the secret $u_1 \in_R Z_q$ is chosen by $\mathbf{U}$ such that $g_1^{u_1} g_2 \neq 1$. It is needed for $\mathbf{U}$ to prove to $\mathbf{B}$ that he knows the represent of I with respect to g_1. $\mathbf{U}$ computes $z' = h_1^{u_1} h_2 = (I g_2)^x$.

Withdrawal: $\mathbf{U}$ gets a coin $(A, B, (z, a, b, r))$ using blind signature.

1. $\mathbf{U}$ proves to $\mathbf{B}$ that he is the owner of some account.
2. $\mathbf{B}$ chooses randomly and uniformly $w \in_R Z_q$, compute $a' = g^w$, $b' = (I g_2)^w$, and then sends to $\mathbf{U}$ a' and b'.
3. $\mathbf{U}$ selects $s, x_1, x_2, u, v \in_R Z_q$; computes $A = (I g_2)^s$, $z = (z')^s$, $B_1 = g_1^{x_1}$, $B_2 = g_2^{x_2}$, $B = [B_1, B_2]$, $a = (a')^u g^v$, $b = (b')^{su} A^v$, $c = H(A, B, z, a, b)$, $c' = c/u$; and then sends c' to $\mathbf{B}$.
4. $\mathbf{B}$ responses $r' = c'x + w$ to $\mathbf{U}$.
5. $\mathbf{U}$ calculates $r = r'u + v$, checks $g^{r'} = h^{c'} a'$ and $(I g_2)^{r'} = (z')^{c'} b'$. $\mathbf{U}$ obtains a coin $(A, B, Sig_{Bank}(A, B))$ where $Sig_{Bank}(A, B) = (z, a, b, r)$, such that $g^r = h^{H(A,B,z,a,b)}$ and $A^r = z^{H(A,B,z,a,b)} b$.

Payment:

1. $\mathbf{U}$ sends to $\mathbf{S}$: $A_1 = g_1^{u_1 s}$, $A_2 = g_2^s$, $A, B, (z, a, b, r)$.
2. $\mathbf{S}$ checks $A = A_1 A_2$, $A \neq 1$, $sig(A, B) = (z, a, b, r)$. If all these succeed, $\mathbf{S}$ sends $\mathbf{U}$: $d = H_1(A_1, B_1, A_2, B_2, I_S, date/time)$.
3. $\mathbf{U}$ computes $r_1 = d(u_1 s) + x_1$, $r_2 = ds + x_2$, and then sends r_1, r_2 to $\mathbf{S}$.
4. $\mathbf{S}$ accepts iff $g_1^{r_1} = A_1^d B_1$ and $g_2^{r_2} = A_2^d B_2$.

Deposit:

1. $\mathbf{S}$ sends $\mathbf{B}$ the transcript of payment A_1^i, B_1^i, A_2^i, B_2^i, (z^i, a^i, b^i, r^i), I_S^i, $date/time^i$, d^i, r_1^i, r_2^i for $i = 1, \cdots, n$.
2. $\mathbf{B}$ checks I_S^i is the identity of the shop, and does whatever $\mathbf{S}$ has done in payment protocol for $i = 1, \cdots, n$. Finally $\mathbf{B}$'s account is credited.

Appendix 2   The Protocols of [JY96, J97]

All the notations used here are as in [JY96, J97]. Briefly, (S, V) is some existentially unforgeable signature scheme. A coin is a pair (y, s) where $s = S_B(S_O(y))$ and y is the public key corresponding to the secret key x.

Withdrawal:

1. **U** runs the key generation algorithm to get (x, y). He proves his identity to **B** potentially in a manner that does not allow an eavesdropper to get any information about his identity.
2. Using the magic ink signature generation scheme, **U**, **B**, and the Ombudsman servers compute an output so that the withdrawer gets a **B/O** signature $s = S_B(S_O(y))$, and **B** (and possibly the **O**) server gets a tag tag linked to the signed message, i.e., such that $Corresponds(tag, coin)$ is satisfied.

Payment:

1. **U** sends **S**: (y, s).
2. **S** checks $s = S_B(S_O(y))$ and sends back challenge c to **U**.
3. **U** sends **S** the answer $a = S_x(c)$.
4. **S** checks $V_y(a, c) = 1$ and keeps the transcript (y, s, c, a).

Deposit:

1. **S** forwards to **B** the transcript (y^i, s^i, c^i, a^i) where $i = 1, \cdots, n$.
2. **B** checks that $s^i = S_B(S_O(y^i))$ and $V_{y^i}(a^i, c^i) = 1$, and further verifies that the same transcript has not been deposited before, and then credits the depositor's account.

Engineering an eCash System

Tim Ebringer and Peter Thorne

University of Melbourne, Parkville Victoria, 3052, Australia,
{tde,pgt}@cs.mu.oz.au,
http://www.cs.mu.oz.au/~tde

Abstract. Since the seminal eCash paper by Chaum, Fiat and Naor [4], there have been many improvements to the original protocol. However, a protocol constitutes only a single layer of a production system. We examine the feasibility of using some of these new eCash protocols in a more massive system that replaces our current physical cash system. Our examples are taken from the Australian economy, but should be applicable to other developed nations.

Introduction

Cash is something of an ugly duckling in the payment systems architecture. It retains some unusual properties that have proven to be rather difficult to emulate electronically. However, it remains a crucial component of the available payment systems.

Ever since it was recognised that it might be possible to build an eCash infrastructure which could, theoretically at least, replace our current paper and coin system, there has been a wealth of research in the area. However, the outcome of much of this work has been variations on the initial protocol, with little sense of an ultimate goal in creating a system. 'Efficiency' has been a major theme, but there has been limited justification for exactly where efficiency optimisations should be applied.

eCash is an attempt to marry the 'difficult to emulate' features of *cash* with the efficiency and cost benefits gleaned from an *electronic* payments system.

The Importance of Cash Experience and anecdotal evidence within Australia indicate that cash accounts for the largest quantity of transactions, but only a fraction of their value [3]. Data to verify this, however, is not available. Payments are measured as a flow — the value or volume during a period — while most measures of cash are of its stock at a point in time. A number of proxies are therefore used to give an indication of the role of cash.

The most commonly used proxy is the stock of cash in circulation. Current market indicators show that cash usage is by no means on the decline. Rather, usage of cash appears to have increased in recent years.

Each month, Australians currently make a little over 30 million withdrawals from bank ATMs; these average around $150 per withdrawal and have a total value of about $5 billion per month. The number of cash withdrawals has grown

M. Mambo, Y. Zheng (Eds.): ISW'99, LNCS 1729, pp. 32–36, 1999.
© Springer-Verlag Berlin Heidelberg 1999

fairly slowly over the past few years but the value of withdrawals has risen by around 40 per cent, implying a significant increase in the average amount withdrawn. This may reflect an increase in both the demand for cash to make payments and in average holdings of cash. There is anecdotal evidence that customers are economising on trips to bank branches and/or the ATM to minimise bank fees. The current low-inflation environment may also have induced consumers to hold larger amounts of cash.

Although ATM withdrawals do not give any indication of the total value of cash transactions, they do give a lower bound. With annual cash withdrawals of just under $60 billion, withdrawals from bank ATMs alone still exceed the value of payments made by EFTPOS and credit cards combined (see below). The total value of cash transactions would therefore well outstrip transactions using these payment instruments. Cash remains an important payment instrument in the Australian economy.

The Risks and Rewards of eCash Despite the rather chaotic development of eCash, there is a compelling rationale behind it. Firstly, cash is expensive. Estimates vary widely as to the exact cost of cash since it is probably highly dependent upon the host country, but estimates have placed the cost per transaction from between $0.015 U.S. (A$0.015) in 1975 to sFr$ 0.83 (A$0.95) in 1984. The Reserve Bank of Australia alone spent $22 million on storage of cash in 1998 [11]. eCash holds the promise of alleviating the problems of printing, storing and distributing cash.

Furthermore, an electronic implementation of cash lends certain flexibilities to its use. For example, eCash should allow the user to keep careful track of their money. If engineered well, it should be possible for developers who are creating devices to manage eCash to incorporate sophisticated enhancements to the basic protocol. Such enhancements might include an ability to generate electronic receipts for tax or audit purposes.

There are, however, considerable risks in developing an eCash system. Firstly, there is significant resistance to its adoption by the general public. This is probably due to users' distrust of software in general. Many consumers imagine their money disappearing as easily as their documents to a certain infamous 'blue screen of death'. Users of cash feel comfortable that they have something tangible which is worth value. It is impossible to duplicate this concept electronically, and this will certainly hinder the acceptance of eCash.

The cash economy also presents an attractive target for sabotage. Crippling a nation's eCash economy would be a serious blow in any confrontation. It is unlikely, at least until a nation as a whole is completely at ease with the information age, that eCash will completely replace cash.

Existing Requirements eCash was originally proposed by Chaum, Fiat and Naor [4]. Since then, eCash has been a popular area in which to demonstrate cryptographic techniques. There has been considerable work in theoretical designs of eCash, notably in improving its efficiency, making the coins divisible [6], proving the security and more recently in the more politically sensitive area of anonymity control.

Fukisaki and Okamoto proposed that an eCash protocol should exhibit certain technical characteristics [9] which include security, off-line capabilities, divisibility, transferability and privacy with escrow.

Engineering Practical eCash Systems to Replace Cash

In Australia, less than half of all households have a computer, and furthermore only 57% of these have a modem [1]. Thus, a completely on-line eCash system will not suffice. The system must allow for the bulk of the user/merchant transactions to be conducted off-line. However, if a system is off-line, it is impossible to prevent double-spending coins — you can only detect it afterwards. Brands suggests using tamper-resistant hardware as an "observer" during a transaction [2], but whether such hardware will render the cost of breaking a system uneconomic will depend on the actual implementation.

Another difficulty is the problem of transferability. This feature significantly complicates the system. We argue that if the system possesses divisibility, then transferability, whilst desirable, is not essential. Ordinary cash is transferable, but not divisible. The fact that ordinary money is transferable makes up for the fact that we may not have exact change in our wallets since we may exchange coins of different denominations. If we were to carry a single divisible $150 note, we would *always* be able to make up an exact amount, provided we had enough money.

We propose that the Okamoto/Fujisaki requirements be extended with some engineering requirements such as dependability, useability, expiry and careful distribution of secrets. Dependability encompasses availability, reliability, safety and integrity. Useability is crucial to widespread consumer acceptance of technology, and is where some smartcard systems have failed in the past [10]. Expiry of coins must be enforced so that the banks need not keep a record of every coin for all eternity. Finally, if an escrow type system is used, the escrow agencies must have the confidence of the user community.

Engineering and the Marketplace

Anonymity and Culture Probably the attribute of eCash that has the most disparity in acceptance between cultures is that of anonymity. The relationship between government and constituency and the amount of trust involved differs widely. Australians, for example, are typically skeptical of their politicians' good intentions. Furthermore, studies have shown that anonymity is easy to abuse [12].

The solution is to use some form of escrow. That is, at the time of coin generation, the user must lodge their identity amongst some trustees using a threshold scheme. Who the trustees are, and what the threshold is, must be carefully chosen. We propose at least one civilian and one governmental trustee. For example, in Australia we might use a (3,2) threshold scheme with the trustees being the Reserve Bank, the Law Institute and the Attorney General's Office, with the Attorney General only being consulted in the case of a deadlock between the other two agencies.

Technological Infrastructure An eCash system has certain minimal technological requirements. Increasing the robustness of the system tends to escalate costs. However, a robust eCash system is crucial, since it is an essential component of the economy.

The infrastructure required can be estimated from the number of users, the transaction density by geographic area, the data storage requirements, the bandwidth requirements and the computation required per transaction. This system is extremely complicated to model, but we shall contrive an example to estimate the order we are dealing with.

If we take Sydney, where we will assume data telecommunication abilities to be readily accessible, then in 1994 there were 3.7 million people living there [8]. If we conservatively estimate everyone to make less than 10 eCash transactions in a day, and we also assume that a $150 coin will last for at least 2 days, then we will need to be able to generate around 1.9 million coins per day. We will also need to process about 38 million deposits per day.

To process the expected number of transactions from Sydney alone, we would need a database of considerable power. The most powerful database on the TPC-C[1] list runs Oracle 8 on an Alphaserver 8400/8. This machine can process 10^5 transactions of the TPC-C type per minute (and costs over \$US 14 million). Thus, it could do 38 million transaction in about 6.8 hours.

This in no way takes into account the (usually computationally expensive) cryptographic operations needed to complement the transaction. This alone would demand a set of massively parallel cryptographic coprocessors.

The crucial lesson from this is that it is unlikely that a single database will be able to deal with the expected volume of transactions. Rather, a more 'distributed' approach is called for: coins must be marked with the database and trustee nodes that were used in their generation so that the deposit can be completed.

Geographic Considerations Countries that have very rugged terrain, isolated communities and/or large distances to contend with will face difficulties in mustering the communications systems needed to implement an eCash system.

Several options have been proposed for rural communication: firstly, in Australia today, most people in the outback use a radio for communication. This is easily adaptable to a packet radio system that can transmit data at 4800 baud. This should be sufficient for a small volume of transactions. Satellite has also been suggested. Satellite bandwidth is extremely expensive, and can probably only be used as a last resort.

Any one of these systems may be sufficient, depending on the exact nature of the protocol employed, yet rural communities would have little to gain from

---

[1] Transaction Processing Performance Council (see http://www.tpc.org)

an eCash system anyway. Any eCash system will, at least for a while, have to exist in tandem with the old physical system.

Conclusion

The cryptographic knowledge required to replace a cash infrastructure with eCash is already available and the appropriate communications infrastructure already exists in developed nations. However, consumer acceptance and the significant computational load that issuers must support remain as barriers that have yet to be overcome.

Acknowledgements We would like to thank Walter Neumann, Yuliang Zheng and Ed Kazmierczak. Walter for his invaluable and lucid assistance with number theory; Yuliang for his enthusiastic support and advice regarding smartcards, wallets and cryptography in general; and finally Ed for his finely honed criticism and skill in reading and commenting on the drafts.

References

1. Australian Bureau of Statistics. *8146.0 Household Use of Information Technology, Australia*, 1998. ISSN: 1329-4067.
2. S. Brands. Untraceable off-line cash in wallets with observers. In *Advances in Cryptology — Crypto '93, Proceedings (Lecture Notes in Computer Science 773)*, pages 302–318. Springer-Verlag, 1993.
3. Michele Bullock and Luci Ellis. Some features of the Australian payments system. Reserve Bank of Australia, December 1998.
4. D. Chaum, A. Fiat, and M. Naor. Untraceable electronic cash. In *Advances in Cryptology —Crypto '88 (Lecture Notes in Computer Science)*, pages 319–327. Springer-Verlag, 1990.
5. D. Chaum and T.P. Pedersen. Transferred cash grows in size. In *Advances in Cryptology — Eurocrypt '92, Proceedings (Lecture Notes in Computer Science 658)*, pages 390–407. Springer-Verlag, 1993.
6. T. Eng and T. Okamoto. Single-term divisible electronic coins. In *Advances in Cryptology — Eurocrypt '94, Proceedings*, pages 306 – 319, New York, 1994. Springer-Verlag.
7. Jean-Paul Boly et al. The esprit project cafe. In *ESORICS '94 Proceedings*, pages 217–230. Springer-Verlag, 1994.
8. Jing Shu et al. *Australia's population trends and prospects, 1995*. Canberra : Australian Govt. Pub. Service, 1996, 1996.
9. Eiichiro Fujisaki and Tatsuaki Okamoto. Practical escrow cash systems. In Mark Lomas, editor, *Security Protocols : International Workshop (Lecture Notes in Computer Science 1189)*, pages 33–47. Springer-Verlag, 1997. Cambridge, United Kingdom, April 10-12.
10. *NIKKEI Digital Money Systems*, 15 October 1998.
11. Profit and earnings distribution. Reserve Bank of Australia, 1998.
12. B. von Solms and D. Naccache. On blind signatures and perfect crimes. *Computers and Security*, 11(6):581–583, October 1992.

Unlinkable Electronic Coupon Protocol with Anonymity Control

Toru Nakanishi, Nobuaki Haruna, and Yuji Sugiyama

Department of Information Technology,
Faculty of Engineering, Okayama University,
3-1 tsushimanaka, Okayama 700-8530, Japan
nakanisi@in.it.okayama-u.ac.jp

Abstract. In a variant of the electronic cash protocol, an electronic coupon protocol, a withdrawn coin is divided into many sub-coins whose face values are fixed in advance, and the sub-coins are only used in payments. The original coin is called ticket and the sub-coins are called sub-tickets. The electronic cash protocol should satisfy not only the anonymity that the payer cannot be traced from the payments, but also the unlinkability. The unlinkability means that anyone cannot determine whether payments were made by the same payer. If the unlinkability does not hold, tracing the payer from one payment leads to tracing the payer from all his/her payments, and the link between the payments also facilitates the de-anonymization. In the previously proposed electronic coupon protocol, payments of sub-tickets derived from the same ticket are linkable. Since the complete anonymity of payments facilitates fraud and criminal acts, the electronic cash protocols should equip the revocation of the anonymity. In this paper, an electronic coupon protocol is proposed, where all payments are unlinkable but the anonymity of the payments can be revoked.

1 Introduction

In electronic cash protocols, the privacy of customers should be protected, where not only the anonymity holds but the unlinkability also should hold. The anonymity means that no other party except trustees can trace the payer from the transcript of his/her payment, and the unlinkability means that no other party except trustees can determine whether any pair of the transcripts are made by the same payer. Consider the case that the anonymity holds and the unlinkability does not hold. Then, if a party can trace the payer from a transcript by any other means, the party can also trace all transcripts of the payer. In the addition to this, it facilitates de-anonymization [1], that is, given the history of linkable transcripts of an anonymous payer, a party may compare the history with the bank's information about when and how much money each real person withdrew, and thus may trace the payer.

A variant of the electronic cash protocol, an *electronic coupon protocol*, is proposed [2], where a withdrawn coin is divided into many sub-coins whose face

M. Mambo, Y. Zheng (Eds.): ISW'99, LNCS 1729, pp. 37–46, 1999.

values are fixed in advance and the summation of the face values of all sub-coins is equal to the face value of the original coin. Note that the only sub-coins can be spent and the original coin cannot be spent. The original coin is called *ticket* and the sub-coins are called *sub-tickets*. For example, a customer withdraws a ticket with \$100, divides ten sub-tickets with \$10, and pays the sub-tickets. In the electronic coupon protocol proposed in [2], a transcript of a payment contains a withdrawn ticket itself which is a digital signature to assure the validity of the ticket. Thus, the transcripts of sub-tickets derived from the same ticket are linkable though the transcripts of sub-tickets derived from the different tickets are unlinkable.

In this paper, an electronic coupon protocol is proposed, where any pair of the transcripts are unlinkable. The proposed protocol uses a group signature scheme proposed in [3]. The *group signature scheme* satisfies the followings:

1. Only members of the group can sign messages, and anyone can determine whether a given signature is a valid signature of a member of the group.
2. Any other party except trustees cannot find out which group member signed a message nor determine whether two signatures have been issued by the same group member.
3. In the case of a subsequent dispute, the signer can be identified.

In the scheme of [3], the group consists of owners of unforgeable certificates issued from the group manager. If the certificate is used as a ticket and the group signature is used as a transcript of a payment, the followings holds: The property 1 of the group signature scheme assures that a payer has a ticket, the property 2 assures that the payment is anonymous and unlinkable, and the property 3 enables an over-spending payer to be identified if over-spending can be detected. However, if the transcript of the payment is only the group signature, the detection of over-spending is infeasible since the payments of the same sub-ticket are unlinkable. In the protocol of this paper, the transcript of the payment consists of the group signature and a value which is the same if and only if the payer uses the same sub-ticket. Thus, over-spending is detected with the valid transcripts of payments unlinkable.

Though the electronic cash protocol requires the anonymity, the complete anonymity facilitates fraud and criminal acts, such as money laundering [4], anonymous blackmailing and illegal purchases. Thus, the protocols should equip the revocation of the anonymity [5]. The revocation must be accomplished selectively, that is, only the transcript of the payment for which a judge's order is given must be de-anonymized. There are two type of revocation procedures, owner tracing to identify the payer of a payment, and coin tracing to link the withdrawal of a coin to the payments of the coin. Owing to the owner tracing, a person who conducted an illegal purchase can be identified. Owing to the coin tracing, payments of a coin obtained illegally can be traced. The protocol proposed in this paper has both owner and coin tracings where the payments of unrelated customers remains anonymous.

This paper is organized as follows: Section 2 describes requirements for the electronic coupon protocols. Section 3.1, where notations are shown, is followed by Section 3.2, where an electronic coupon protocol satisfying the requirements is proposed. Section 3.3 discusses the security of the proposed protocol. Section 4 concludes this paper.

2   Requirements for Electronic Coupon Protocols

The requirements for electronic coupon protocols are as follows [2], [5]:

Unforgeability: A ticket and a transcript of a payment can not be forged.

Unreusability: The customer who spends a sub-ticket twice or more can be identified.

No swindling: No one except the customer who withdraws a ticket can spend a sub-ticket derived from the ticket. The deposit information can not be forged.

Anonymity: No one except the payer and the trustees can trace the payer from the payment.

Unlinkability: No one except the payer and the trustees can determine whether any pair of payments is executed by the same customer.

Anonymity Revocation: Anonymity of a transcript of a payment can be revoked only by the trustees and when necessary, where the following revocation procedures should be accomplished:

Owner Tracing: To identify the payer of a payment.

Ticket Tracing(Coin Tracing): To link the withdrawal of a ticket to the payments derived from the ticket.

Only the transcript for which a judge's order is given must be de-anonymized.

Divisibility: A ticket is divided into many sub-tickets whose face values are fixed in advance and the summation of the face values of all sub-tickets is equal to the face value of the original ticket.

Off-line-ness: During payments, the payer communicates with only the shop.

3   Electronic Coupon Protocol

3.1   Preliminaries

As mentioned in Introduction, the proposed protocol uses the group signature scheme of [3]. This subsection reviews notations and primitives used in [3]. Let $\tilde{0}$ be the empty string. If A is a set, $a \in_R A$ means that a is chosen at random from A according to the uniform distribution. Let $G = \langle g \rangle$ be a cyclic group of order n, where g is a generator of G. For example, G could be a subgroup of Z_r^*, for a prime r with $n|(r-1)$. The discrete logarithm of $y \in G$ to the base g is the smallest positive integer x satisfying $g^x = y$. Similarly, the double discrete logarithm of $y \in G$ to the bases g and h is the smallest positive integer x satisfying $g^{(h^x)} = y$ if such an x exists. The parameters $n, G, g,$ and h should be chosen such that computing discrete logarithms in G to bases g and h is infeasible. An e-th root of the discrete logarithm of $y \in G$ to the base g is an integer x satisfying $g^{(x^e)} = y$ if such an x exists. Assume that computing e-th roots in Z_n^* is infeasible if the factorization of n is unknown.

As a primitive to prove the knowledge of secret values without leaking any useful information on the secret, the signature of the knowledge is used. That is a non-interactive proof system and a signature on a message, that is, only one who knows secret values satisfying a statement can compute the signature, a verifier cannot obtain any useful information about the secret values, and an adversary cannot compute the signature on an unsigned message by using signatures on messages chosen by the adversary. One of the signatures of the knowledge is the signature on a message m of an entity knowing the discrete logarithm x of y, which is basically a Schnorr signature [6]. The signatures of the knowledge of values satisfying more complex statement can be also constructed. Four types of signatures of the knowledge are used to construct the proposed electronic coupon protocol. Three signatures of the knowledge are shown in [3] and the other is shown in [7]. The first one is the signature of the knowledge of representations of $y_1, \ldots, y_w$ to the bases $g_1, \ldots, g_v$ on message m, and it is denoted as

$$SKREP[(\alpha_1, \ldots, \alpha_u) : (y_1 = \prod_{j=1}^{l_1} g_{b_{1j}}^{\alpha_{e_{1j}}}) \wedge \cdots \wedge (y_w = \prod_{j=1}^{l_w} g_{b_{wj}}^{\alpha_{e_{wj}}})](m),$$

where constants $l_i \in \{1, \ldots v\}$ indicate the number of bases on representation of y_i, the indices $e_{ij} \in \{1, \ldots, u\}$ refer to the elements $\alpha_1, \ldots, \alpha_u$ and the indices $b_{ij} \in \{1, \ldots, v\}$ refer to the elements $g_1, \ldots, g_v$. For example, $SKREP[(\alpha, \beta) : y = g^\alpha \wedge z = g^\beta h^\alpha](m)$ is the signature on m of an entity knowing the discrete logarithm of y to the base g and a representation of z to the bases g and h, where the h-part of this representation equals the discrete logarithm of y to the base g. The second type is the signature of the knowledge of the e-th root of the discrete logarithm of y to the base g on m, and is denoted as

$$E - SKROOTLOG[\beta : y = g^{\beta^e}](m).$$

The third type is the signature of the knowledge of the e-th root of the g-part of a representation of y to the bases h and g on m, and is denoted as

$$E - SKROOTREP[(\gamma, \delta) : y = h^\gamma g^{\delta^e}](m).$$

For the constructions of these signatures, refer to [3]. Note that E-$SKROOTLOG$ and $E - SKROOTREP$ are efficient if e is small. The last type, which is shown in [7], is the signature of the knowledge of the discrete logarithm of y to the base g and the double discrete logarithm of z to the bases g and h on m, where the discrete logarithm of y to the base g equals the double discrete logarithm of z to the bases g and h. This is denoted as

$$SKLOGLOGREP[\epsilon : y = g^\epsilon \wedge z = g^{(h^\epsilon)}](m).$$

Note that there is a difference between $SKLOGLOGREP$ in this paper and that in [7]. The difference is the order of the group. The order in this paper is n which is not a prime, but that in [7] is a prime, which does not affect the proof that SKLOGLOGREP is the signature of the knowledge.

3.2 Protocol

The participants in the proposed protocol are customers, a bank, shops, and a trustee. The proposed protocol consists of seven sub-protocols, that is, the setup, withdrawal, payment, deposit, over-spender tracing, owner tracing, and ticket tracing sub-protocols. In the setup sub-protocol, all participants set up the electronic coupon protocol. The withdrawal sub-protocol is executed when a customer withdraws a ticket from the bank. By using the payment sub-protocol, the customer pays a shop a sub-ticket derived from the ticket. In the deposit sub-protocol, the shop sends the bank the transcript of the payment to deposit the paid sub-ticket in the bank. If, at this time, over-spending the sub-ticket is detected, the bank with the cooperation of the trustee identifies the over-spender in the over-spender tracing sub-protocol. As the owner and ticket tracing, the owner and ticket tracing sub-protocols are used.

Next, each sub-protocol is described.

Setup:

Assume that a ticket is divided into k sub-tickets, each sub-ticket is assigned to the index i $(1 \leq i \leq k)$, and all participants agree on the type of ticket, that is, the face values of the ticket and the sub-tickets, and the indices in advance. If different type of tickets are used, the bank and trustee must execute the following Step 1 and 2 to make secret and public keys for each type of ticket. Thus, a public key is assigned to a type of ticket.

1. The bank computes the followings:
   - An RSA modulus n and two public exponent $e_1, e_2 > 1$,
   - Two integers $f_1, f_2 > 1$,
   - A cyclic group $G = \langle g \rangle$ of order n in which computing discrete logarithms is infeasible,
   - An element $h \in G$ whose discrete logarithm to base g is unknown.

   The public key for the bank is $\mathcal{Y} = (n, e_1, e_2, f_1, f_2, G, g, h)$, and the secret key is the factorization of n. Note that e_1, e_2, f_1 and f_2 must satisfy that solving the congruence $f_1 x^{e_1} + f_2 \equiv v^{e_2} \pmod{n}$ is infeasible. The choices for e_1, e_2, f_1 and f_2 are discussed in [3].

2. For i $(1 \leq i \leq k)$, the trustee chooses $t_i, \rho \in_R Z_n^*$ to compute $h_i = g^{t_i}$ and $y_R = h^\rho$. The trustee makes h_i and y_R public, and keeps t_i and ρ secret. Note that h_i is used in the payment of i-th sub-ticket.

3. Each participant publishes the public key of any digital signature scheme and keeps the corresponding secret key. Hereafter, except in the payment sub-protocol, the values sent from each participant are signed on the digital signature scheme.

Withdrawal:

1. A customer chooses $x \in_R Z_n^*$ to compute $y = x^{e_1} \bmod n$ and $z = g^y$. Then, the customer chooses $r_1, r_2 \in_R Z_n^*$ to compute $\tilde{y} = r_1^{e_2}(f_1 y + f_2) \bmod n$, $C_1 = h^{r_2} h^y$, and $C_2 = y_R^{r_2}$. Furthermore, the customer computes the following signatures of the knowledge:

$$V_1 = E\text{-}SKROOTLOG[\alpha : z = g^{\alpha^{e_1}}](\tilde{0}),$$
$$V_2 = E\text{-}SKROOTLOG[\beta : g^{\tilde{y}} = (z^{f_1}g^{f_2})^{\beta^{e_2}}](\tilde{0}),$$
$$V_3 = SKREP[(\gamma, \delta) : C_1 = h^\gamma h^\delta \wedge C_2 = y_R^\gamma \wedge z = g^\delta](\tilde{0}).$$

V_1 proves that z is the form of $g^{\alpha^{e_1}}$ for α which the customer knows. V_2 proves that $\tilde{y} \equiv \beta^{e_2}(f_1\alpha^{e_1} + f_2) \pmod{n}$ for β which the customer knows. Thus, the correctness of $\tilde{y}$ and z is assured. V_3 assures that (C_1, C_2) is the ElGamal encryption [8] of an element, where the discrete logarithm of the element to the base h equals to that of z to the base g. The customer sends the bank $(\tilde{y}, z, C_1, C_2, V_1, V_2, V_3)$.

2. If V_1, V_2, and V_3 are correct, the bank sends the customer $\tilde{v} = \tilde{y}^{1/e_2} \bmod n$ to charge the customer's account the face value of the ticket.

3. The customer computes $v = \tilde{v}/r_1 \bmod n$ to obtain the ticket (x, v), where $v \equiv (f_1 x^{e_1} + f_2)^{1/e_2} \pmod{n}$.

Payment: Assume that each shop is numbered in order to be identified. Let m be the concatenation between the number of the shop which obtains the payment and the time when the payment is made.

1. When a customer spends i-th sub-ticket, the customer computes $\tilde{g} = g^{\tilde{r}_1}, \tilde{z} = \tilde{g}^y, \tilde{h}_i = h_i^{h^y}, \tilde{C}_1 = h^{\tilde{r}_2}g^y$ and $\tilde{C}_2 = y_R^{\tilde{r}_2}$, where $\tilde{r}_1, \tilde{r}_2 \in_R Z_n^*$. Furthermore, the customer computes the following signatures of knowledge:

$$\tilde{V}_1 = E - SKROOTREP[(\alpha, \beta) : \tilde{C}_1 = h^\alpha g^{\beta^{e_1}}](m),$$
$$\tilde{V}_2 = E - SKROOTREP[(\gamma, \delta) : \tilde{C}_1^{f_1}g^{f_2} = h^\gamma g^{\delta^{e_2}}](m),$$
$$\tilde{V}_3 = SKREP[(\epsilon, \zeta) : \tilde{C}_1 = h^\epsilon g^\zeta \wedge \tilde{C}_2 = y_R^\epsilon](m),$$
$$\tilde{V}_4 = SKREP[(\eta, \theta) : \tilde{z} = \tilde{g}^\eta \wedge \tilde{C}_1 = h^\theta g^\eta](m),$$
$$\tilde{V}_5 = SKLOGLOGREP[\iota : \tilde{z} = \tilde{g}^\iota \wedge \tilde{h}_i = h_i^{h^\iota}](m).$$

$\tilde{V}_1$ proves that $\tilde{C}_1$ is the form of $h^\alpha g^{\beta^{e_1}}$ for α and β which the customer knows. $\tilde{V}_2$ proves that $f_1\alpha \equiv \gamma \pmod{n}$ and $f_1\beta^{e_1}+f_2 \equiv \delta^{e_2} \pmod{n}$ for γ and δ which the customer knows. $\tilde{V}_3$ proves that $(\tilde{C}_1, \tilde{C}_2)$ is the ElGamal encryption of an element in G. $\tilde{V}_4$ and $\tilde{V}_5$ assure that the discrete logarithm of the encrypted element to the base g equals to the double discrete logarithm of $\tilde{h}_i$ to the base h_i and h. The customer sends the shop the transcript of the payment $(i, \tilde{g}, \tilde{z}, \tilde{h}_i, \tilde{C}_1, \tilde{C}_2, \tilde{V}_1, \tilde{V}_2, \tilde{V}_3, \tilde{V}_4, \tilde{V}_5)$.

2. The shop verifies that the transcript is correctly formed.

Deposit:

1. The shop sends the bank the transcript of the payment $(i, \tilde{g}, \tilde{z}, \tilde{h}_i, \tilde{C}_1, \tilde{C}_2, \tilde{V}_1, \tilde{V}_2, \tilde{V}_3, \tilde{V}_4, \tilde{V}_5)$.

2. The bank verifies that the transcript is correctly formed. Then, the bank checks whether the payment causes the sub-ticket to be over-spent by checking whether $(i, \tilde{h}_i)$ occurs in the transcripts of all previously deposited payments. If it occurs, the over-spender tracing sub-protocol is followed since the sub-ticket of the payment is over-spent. Otherwise, the face value of the sub-ticket is deposited in the shop's account, and the transcript of the payment is kept in the bank's database.

Over-spender tracing:

1. The bank sends the trustee two transcripts of the payments where the same sub-ticket is spent, $(i, \tilde{g}, \tilde{z}, \tilde{h}_i, \tilde{C}_1, \tilde{C}_2, \tilde{V}_1, \tilde{V}_2, \tilde{V}_3, \tilde{V}_4, \tilde{V}_5)$ and $(i', \tilde{g}', \tilde{z}', \tilde{h}_i', \tilde{C}_1', \tilde{C}_2', \tilde{V}_1', \tilde{V}_2', \tilde{V}_3', \tilde{V}_4', \tilde{V}_5')$.
2. The trustee verifies that the transcripts are correctly formed. If they are correctly formed, the trustee sends the bank $\hat{z} = \tilde{C}_1/\tilde{C}_2^{1/\rho}$ and $SKREP[\alpha : \tilde{C}_1 = \hat{z}\tilde{C}_2^{\alpha} \wedge h = y_R^{\alpha}](\tilde{0})$. This $SKREP$ proves that $(\tilde{C}_1, \tilde{C}_2)$ is decrypted into $\hat{z}$.
3. The bank searches z identical with $\hat{z}$ to present the customer's signature on z, which indicates the over-spender.

Owner tracing:

Except that the transcript of the targeted payment is sent in Step 1, this sub-protocol is the same as the over-spender tracing sub-protocol.

Ticket tracing:

1. The bank sends the trustee the transcript of the targeted withdrawal $(\tilde{y}, z, C_1, C_2, V_1, V_2, V_3)$.
2. The trustee verifies that the transcript is correctly formed. If it is correctly formed, the trustee sends the bank $\hat{h} = C_1/C_2^{1/\rho}$ and $SKREP[\alpha : C_1 = \hat{h}C_2^{\alpha} \wedge h = y_R^{\alpha}](\tilde{0})$. This $SKREP$ proves that (C_1, C_2) is decrypted into $\hat{h}$.
3. For the sent $\hat{h}$, the bank (and shops) checks whether $\tilde{h}_i = \hat{h}_i^{\hat{h}}$ on a transcript of a payment $(i, \tilde{g}, \tilde{z}, \tilde{h}_i, \tilde{C}_1, \tilde{C}_2, \tilde{V}_1, \tilde{V}_2, \tilde{V}_3, \tilde{V}_4, \tilde{V}_5)$. If it holds, the transcript is derived from the targeted withdrawal.

Since the proposed protocol uses the group signature scheme in [3], the correspondence and differences between the proposed protocol and the group signature scheme are discussed. The setup sub-protocol corresponds to the setup on the original scheme. The withdrawal sub-protocol corresponds to issuing of the membership certificate on the original scheme. The membership certificate is used as a ticket. To send (C_1, C_2) which is the encryption of h^y with the public key of the trustee, and to send the proof of its correctness are added in the sub-protocol. The group signature corresponds to the transcript of the payment, where $i, \tilde{g}, \tilde{z}, \tilde{h}_i, \tilde{V}_4$, and $\tilde{V}_5$ are added to the signature. $(i, \tilde{h}_i)$ is the information to check whether the sub-ticket has been spent before, which have the same value if and only if the same sub-ticket is spent. The over-spender and owner tracing sub-protocols correspond to the identification of the signer on the original scheme. The ticket tracing sub-protocol is newly added. In [3], the group manager not only manages the membership but identifies the signer. But, in this protocol, the power is distributed to the bank and the trustee.

3.3   Discussion

The protocol proposed in this paper as well as the original group signature scheme is based on the infeasibility to compute or compare the discrete logarithms, the security of the ElGamal encryption, and the infeasibility to compute (x, v) satisfying $f_1 x^{e_1} + f_2 \equiv v^{e_2} \pmod{n}$.

It is discussed that the proposed protocol satisfies the requirements in Section 2.

Unforgeability: From the infeasibility to compute (x, v) satisfying $f_1 x^{e_1} + f_2 \equiv v^{e_2} \pmod{n}$, it is infeasible to forge a ticket. From the soundness of the signature of knowledge, it is infeasible to compute the transcript of the payment without a ticket.

Unreusability: Assume that a customer spends a sub-ticket twice, that is, the customer sends two transcripts of the payments whose (i, x, v) are the same. Then, since the signature of the knowledge assume that $\tilde{h}_i = h_i^{h^{x^{e_1}}}$, $(i, \tilde{h}_i)$ of the transcripts are the same. Thus, in the deposit sub-protocol, the sub-ticket of the transcripts is regarded as over-spent. Since the signature of the knowledge also assures that $(\tilde{C}_1, \tilde{C}_2)$ is the ElGamal encryption of z, the bank cooperating with the trustee can identify the over-spender in the over-spender tracing sub-protocol.

No swindling: The blind signature [9] prevents anyone except a customer who withdrew a ticket (x, v) from obtaining the ticket. Because of the signature of the knowledge, a transcript of a payment does not disclose the information about the ticket. Thus, no other party can spend the sub-tickets of a valid customer.

The deposit information is a transcript of a payment. Since the transcript is unforgeable and no other party can spend the sub-tickets of a valid customer, the deposit information cannot be forged.

Unlinkability: In transcripts of payments, the ElGamal encryption $(\tilde{C}_1, \tilde{C}_2)$ does not disclose the information about z, $\tilde{z}$ is unlinkable as well as $\tilde{h}_i$ because of the infeasibility to compare the discrete logarithms, and $\tilde{V}_1, \tilde{V}_2, \tilde{V}_3, \tilde{V}_4$, and $\tilde{V}_5$ do not disclose the information about x and v since they are the signature of the knowledge. Therefore, the transcripts are unlinkable.

Anonymity revocation:

    Owner Tracing: As well as the over-spender tracing, since the signature of the knowledge assures that $(\tilde{C}_1, \tilde{C}_2)$ is the ElGamal encryption of z, the bank cooperating with the trustee can identify the payer from the targeted payment in the owner tracing sub-protocol, where $(\tilde{C}_1, \tilde{C}_2)$ in the other payments are not decrypted and thus the payments remain anonymous.

    Ticket tracing: The signature of the knowledge assures that (C_1, C_2) is the ElGamal encryption of h^y, and that $\tilde{h}_i = h_i^{h^y}$. Thus, the bank and the shops cooperating with the trustee can trace the transcripts of the payments where h^y in the targeted withdrawal is used. Since (C_1, C_2) in the other withdrawals are not decrypted and thus the other payments remain anonymous.

Since the unlinkability holds, the anonymity also holds. From the description of the protocol, it is shown straightforwardly that the divisibility and off-line-ness hold.

In [4], the following perfect crime on the electronic cash protocol using the blind signature is indicated: A kidnaper forces the bank to issue coins with the threat. Then, the kidnaper can spend the coins with complete anonymity. This attack against the protocol proposed in this paper can be protected as follows: If the extortion of tickets occurs, the trustee reveals h^y in all correctly issued tickets by decrypting (C_1, C_2). Then, an additional on-line verification of the trustee is executed in all payments after this. In the additional on-line verification, the trustee checks that $\tilde{h}_i = h_i^{h^y}$ for $\tilde{h}_i$ in a payment and h^y of all correctly issued tickets. The check succeeds if and only if the payment is derived from a correctly issued ticket. If the check fails, the payment is rejected. Thus, this attack can be protected.

4 Conclusion

In this paper, an electronic coupon protocol with unlinkable payments and an anonymity control has been proposed. Let $\mathcal{N}$ be (total face value of a ticket or coin) / (face value of smallest divisible unit). In the proposed electronic coupon protocol, a payment of any value requires $O(\mathcal{N})$ computations since the computations is in proportion to the number of sub-tickets. In [10], an electronic cash protocol where a payment of any value requires only $O(\log \mathcal{N})$ computations is proposed. However, in the protocol of [10], payments derived from the same coin are linkable. An open problem is to propose an electronic cash protocol where a payment of any value requires only $O(\log \mathcal{N})$ computations and all payments are unlinkable.

References

[1] Pfitzmann, B. and Waidner, M.: How to Break and Repair a "Provably Secure" Untraceable Payment System, *Advances in Cryptology — CRYPTO'91*, LNCS 576, Springer–Verlag, pp. 338–350 (1992).

[2] Okamoto, T. and Ohta, K.: One-Time Zero-Knowledge Authentications and Their Applications to Untraceable Electronic Cash, *IEICE Transactions on Fundamentals of Electronics, Communications and Computer Sciences*, Vol. E81-A, No. 1, pp. 2–10 (1998).

[3] Camenisch, J. and Stadler, M.: Efficient Group Signature Schemes for Large Groups, *Advances in Cryptology — CRYPTO'97*, LNCS 1294, Springer–Verlag, pp. 410–424 (1997).

[4] von Solms, S. and Naccache, D.: On Blind Signatures and Perfect Crimes, *Computers and Security*, Vol. 11, No. 6, pp. 581–583 (1992).

[5] Davida, G., Frankel, Y., Tsiounis, Y. and Yung, M.: Anonymity Control in E-Cash Systems, *In the 1st Financial Cryptography Conference* (1997). Available at http://www.ccs.neu.edu/home/yiannis/pubs.html.

[6] Schnorr, C. P.: Efficient signature generation for smart cards, *Advances in cryptology — CRYPTO'89*, LNCS 435, Springer–Verlag, pp. 239–252 (1990).

[7] Stadler, M.: Publicly Verifiable Secret Sharing, *Advances in Cryptology — EUROCRYPT'96*, LNCS 1070, Springer–Verlag, pp. 190–199 (1996).

[8] ElGamal, T.: A public key cryptosystem and a signature scheme based on discrete logarithms, *Advances in Cryptology — CRYPTO'84*, LNCS 196, Springer–Verlag, pp. 10–18 (1985).

[9] Chaum, D.: Blind Signatures for Untraceable Payments, *Advances in Cryptology: Proceedings of CRYPTO '82*, Plenum Press, pp. 199–203 (1983).

[10] Chan, A., Frankel, Y. and Tsiounis, Y.: Easy Come - Easy Go Divisible Cash, *Advances in Cryptology — EUROCRYPTO'98*, LNCS 1403, Springer–Verlag, pp. 561–575 (1998).

On the Security of the Lee-Chang Group Signature Scheme and Its Derivatives
(Extended Abstract)*

Marc Joye[1,**], Narn-Yih Lee[2], and Tzonelih Hwang[3]

[1] Gemplus Card International
34 rue Guynemer, 92447 Issy-les-Moulineaux Cedex, France
marc.joye@gemplus.com
[2] Dept of Applied Foreign Language, Nan-Tai Institute of Technology
Tainan, Taiwan 710, R.O.C.
nylee@mail.ntc.edu.tw
[3] Institute of Information Engineering, National Cheng-Kung University
Tainan, Taiwan, R.O.C.
hwangtl@server2.iie.ncku.edu.tw

Abstract. W.-B. Lee and C.-C. Chang (1998) proposed a very efficient group signature scheme based on the discrete logarithm problem. This scheme was subsequently improved by Y.-M. Tseng and J.-K. Jan (1999) so that the resulting group signatures are unlinkable. In this paper, we show that any obvious attempt to make unlinkable the Lee-Chang signatures would likely fail. More importantly, we show that both the original Lee-Chang signature scheme and its improved version are universally forgeable.

Keywords: Digital signatures, Group signatures, Cryptanalysis.

1   Group Signatures

A *group signature* [1] is a digital signature that allows a group member to sign a message on behalf of the group so that (i) only group members are able to sign on behalf of the group, (ii) it is not possible to determine which group member made the signature, and (iii) in case of disputes, the group authority can "open" the signature to reveal the identity of the signer.

Group signatures can be used to conceal organizational structures, e.g., when a company or a government agency issues a signed statement. They can also be integrated into an electronic cash system in which several banks can securely distribute anonymous and untraceable e-cash. The group property presents then the further advantage to also conceal the identity of the issuing bank [3].

---

* The full paper is available as technical report (LCIS TR-99-7, May 1999) at URL <http://www.ee.tku.edu.tw/~lcis/techreports/1999/TR-99-7.html>.
** Part of this work was performed while the author was with the Dept of Electrical Engineering, Tamkang University, Tamsui, Taiwan 251, R.O.C.

M. Mambo, Y. Zheng (Eds.): ISW'99, LNCS 1729, pp. 47–51, 1999.

2   Lee-Chang Group Signature

We first begin with a brief review of the Lee-Chang group signature scheme and refer to [2] for a thorough description.

For setting up the system, the group authority, T, selects two large primes p and q so that q divides $(p-1)$. He also chooses a generator g of the (multiplicative) subgroup of order q of $GF(p)$. Finally, he chooses a secret key x_T and computes $y_T = g^{x_T} \bmod p$. The group authority makes public p, q, g and y_T. When a user, say U_i, wants to join the group, he first chooses a secret key x_i and computes his corresponding public key $y_i = g^{x_i} \bmod p$. From this public key, the group authority then computes $r_i = g^{-k_i} y_i^{k_i} \bmod p$ and $s_i = k_i - r_i x_T \bmod q$ for a randomly chosen k_i. The membership certificate of user U_i is the pair (r_i, s_i).

Given a membership certificate, any group member is now able to sign a message on behalf of the group. Let m be the message that user U_i wants to sign on behalf of the group and let $h(\cdot)$ be a publicly known one-way hash function. User i first computes $\alpha_i = y_T^{r_i} g^{s_i} \bmod p$. Next, he chooses a random number t, computes $r = \alpha_i^t \bmod p$, and solves for s the congruence $h(m) \equiv rx_i + ts \pmod q$. The group signature on message m is given by (r, s, r_i, s_i).

Anyone can now check the validity of this signature by first computing $\alpha_i = y_T^{r_i} g^{s_i} \bmod p$ and $DH_i = \alpha_i r_i \bmod p$, and then checking whether

$$\alpha_i^{h(m)} \equiv r^s DH_i^r \pmod p \tag{1}$$

holds or not. Moreover, the group authority knows that this signature was issued by user U_i since he knows that the pair (r_i, s_i) corresponds to that user.

3   Unlinkability

The main property of group signatures is that they allow group members to *anonymously* sign on behalf of the group. A related desirable property is that of *unlinkability*. Two different group signatures are unlinkable if no one (but the group authority) is able to decide whether these signatures were issued by the same group member. In the Lee-Chang scheme, the signatures are obviously linkable since each signature generated by user U_i contains the pair (r_i, s_i). The utility of the Lee-Chang scheme appears thus to be somewhat limited in the context of group signatures, as defined in [1].

4   Tseng-Jan Improved Group Signature

Making the Lee-Chang group signature scheme unlinkable seems easy: it suffices not to send (r_i, s_i) and to modify the signature accordingly. This was first proposed and realized by Tseng and Jan [5].

In their scheme, the system setup is the same as in the Lee-Chang scheme (see Section 2). To sign a message m, user U_i with membership certificate (r_i, s_i) randomly chooses a, b and computes the five quantities $A = r_i^a \bmod p$, $B =$

$s_i - b \bmod q$, $C = r_i\, a \bmod q$, $D = g^a \bmod p$ and $E = g^{ab} \bmod p$. He then computes $\alpha_i = D^B\, y_T{}^C\, E \bmod p$ and $r = \alpha_i{}^t \bmod p$ for a randomly chosen t. He finally solves for s the congruence $h(m) \equiv rx_i + ts \pmod{q}$. The group signature on message m is (r, s, A, B, C, D, E). This signature can be validated by verifying that

$$\alpha_i{}^{h(m)} \equiv DH_i{}^r\, r^s \pmod{p} \tag{2}$$

with $\alpha_i = D^B\, y_T{}^C\, E \bmod p$ and $DH_i = \alpha_i\, A \bmod p$. Finally, in case of disputes, the group authority can recover which member issued a given signature by checking for which value of k_i (and thus for which user U_i, see Section 2) the congruence $D^{k_i} \equiv D^B\, y_T{}^C\, E \pmod{p}$ is satisfied.

5 Cryptanalysis

As already remarked by Lee and Chang [2, Section 2], it is worth observing that the membership certificate (r_i, s_i) of user U_i is a Nyberg-Rueppel signature [4] on "message" DH_i ($\equiv y_i{}^{k_i} \equiv y_T{}^{r_i}\, g^{s_i}\, r_i \pmod{p}$). We analyze the implications in the next two paragraphs. Next, in §5.3, we will show how to universally forge a valid group signature without membership certificate.

5.1 Lee-Chang Signature

Let us first explain how given the membership certificate (r_i, s_i),[1] an adversary can create a new membership certificate (r'_i, s'_i) in the Lee-Chang scheme for a chosen "message" μ. The adversary first recovers DH_i as $DH_i = \alpha_i\, r_i \bmod p$ where $\alpha_i = y_T{}^{r_i}\, g^{s_i} \bmod p$. Next, for a chosen μ, he applies Chinese remaindering to $r'_{i,p} = \alpha_i{}^{-1}\,\mu \bmod p$ and $r'_{i,q} = r_i \bmod q$ to obtain r'_i as

$$r'_i = r'_{i,p} + p\left[p^{-1}(r'_{i,q} - r'_{i,p}) \bmod q\right] . \tag{3}$$

(Note that $r'_i > p$.) He also sets $s'_i = s_i$. One can easily see that, *provided that the range conditions are not checked*, (r'_i, s'_i) will be accepted as a valid Nyberg-Rueppel signature on message μ. Indeed, since $r'_i \equiv r_i \pmod{q}$ and $s'_i = s_i$, it follows that $\alpha'_i :\equiv y_T{}^{r'_i}\, g^{s'_i} \equiv y_T{}^{r_i}\, g^{s_i} \equiv \alpha_i \pmod{p}$, and therefore, noting that $r'_i \equiv \alpha_i{}^{-1}\,\mu \pmod{p}$, (r'_i, s'_i) is the Nyberg-Rueppel signature on $DH'_i :\equiv \alpha'_i\, r'_i \equiv \alpha_i\, \alpha_i{}^{-1}\,\mu \equiv \mu \pmod{p}$. If we now take a closer look at the verification equation in the Lee-Chang signature (i.e., Eq. (1)), we see that if μ is chosen as a power of α_i, i.e., $\mu = \alpha_i{}^w \bmod p$ for a randomly chosen w, then solving for s the congruence $h(m) \equiv st + wr \pmod{q}$ —where as in the original scheme $r = \alpha_i{}^t \bmod p$ for a randomly chosen t— yields a valid group signature (r, s, r'_i, s'_i).

5.2 Tseng-Jan Signature

We now return to the Tseng-Jan group signature scheme. This scheme may be seen as the Lee-Chang scheme where user U_i sends a "randomized" membership

---

[1] We note that (r_i, s_i) is publicly available from any given group signature.

certificate (A, B, C, D, E) instead of (r_i, s_i). But since user U_i knows the pair (r_i, s_i), he can, from it, construct a new pair (r'_i, s'_i) and use it to generate group signatures. This straight-forward application of the previously described forgery, however, does not work: the group authority will still be able to open the signature. But, as we can see, the process can be easily randomized:

(1) Randomly choose v $(\not\equiv 1 \pmod q)$ and compute $\hat{\alpha}_i := y_T^{r_i} g^{s_i} \bmod p$ and $\alpha_i = (\hat{\alpha}_i^v)^a \bmod p$ for a randomly chosen a;
(2) Randomly choose w and set $\mu = (\hat{\alpha}_i^v)^w \bmod p$;
(3) Set $r'_{i,p} = \hat{\alpha}_i^{-v} \mu \bmod p$ and $r'_{i,q} = r_i v \bmod q$ and apply Chinese remaindering to $r'_{i,p}$ and $r'_{i,q}$ to obtain r'_i (see Eq. (3));
(4) Set $s'_i = s_i v \bmod q$;
(5) Choose a random b and compute $A = r'^a_i \bmod p$, $B = s'_i - b \bmod q$, $C = r'_i a \bmod q$, $D = g^a \bmod p$ and $E = g^{ab} \bmod p$;
(6) Compute $r = \alpha_i^t \bmod p$ for a randomly chosen t;
(7) Solve for s the congruence $h(m) \equiv st + wr \pmod q$;
(8) The Tseng-Jan group signature on message m is given by (r, s, A, B, C, D, E).

This group signature cannot be opened since $D^B y_T^C E \equiv (D^{k_i})^v \not\equiv D^{k_i}$ $\pmod p$ (see Section 4). User U_i is thus able to produce *untraceable* (yet valid) group signatures. This forgery differs from the forgery against the Lee-Chang scheme in that it can only be mounted by group members. Furthermore, whereas the forgery against the Lee-Chang scheme is easily avoidable,[2] thwarting the previous attack in the Tseng-Jan scheme is not so obvious. The main problem is due to the fact that a malicious group member can "play" with his membership certificate, (r_i, s_i), to generate a fake (r'_i, s'_i). Moreover, since the values of r'_i and s'_i do not appear explicitly in the resulting group signatures, this malicious member may choose out of range values for r'_i and s'_i. This also means that making unlinkable the Lee-Chang scheme is certainly not an easy issue.

5.3   Universal Forgeries

In §5.1, we have shown that if the range conditions for (r_i, s_i) are not checked, anyone (not necessarily group member) can generate a Lee-Chang group signature which will be validated. In this paragraph, we show that, even if the range conditions are satisfied, it is still possible to generate a valid group signature in the Lee-Chang scheme. Because the attack we present can be mounted equally well by a group member or by a non-group member, we call it *universal forgery*.

To forge a Lee-Chang group signature on a message m, we must find (r, s, r_i, s_i) such that (see Eq. (1))

$$\alpha_i^{h(m)} \equiv r^s DH_i^r \pmod p$$

with $\alpha_i = y_T^{r_i} g^{s_i} \bmod p$ and $DH_i = \alpha_i r_i \bmod p$. In this congruence, the values of α_i and DH_i are imposed —i.e., computed by the verifier from (r_i, s_i)— and

---

[2] It suffices to check that the membership certificate (r_i, s_i) sent along with the group signature satisfies the range conditions $1 \leq r_i < p$ and $0 \leq s_i < q$.

so the apparent intractability to forge signatures. Suppose now that the representations of r and DH_i with respect to bases y_T and g are known. So let $r = y_T{}^v g^w \bmod p$ and $r_i = y_T{}^{v_i} g^{w_i} \bmod p$ (and hence $DH_i \equiv \alpha_i r_i \equiv y_T{}^{r_i+v_i} g^{s_i+w_i} \pmod p$) for some known v, w, v_i, w_i. The above congruence then becomes

$$(y_T{}^{r_i} g^{s_i})^{h(m)} \equiv (y_T{}^v g^w)^s (y_T{}^{r_i+v_i} g^{s_i+w_i})^r \pmod p, \tag{4}$$

which is satisfied whenever

$$\begin{cases} r_i\, h(m) \equiv vs + (r_i + v_i)r \pmod q \\ s_i\, h(m) \equiv ws + (s_i + w_i)r \pmod q \end{cases} . \tag{5}$$

Therefore, an adversary can produce a valid Lee-Chang signature by carrying out the following steps:

(1) Randomly choose v, w, v_i and w_i;
(2) Compute $r = y_T{}^v g^w \bmod p$ and $r_i = y_T{}^{v_i} g^{w_i} \bmod p$;
(3) Solve for s the congruence $r_i\, h(m) \equiv vs + (r_i + v_i)r \pmod q$;
(4) Solve for s_i the congruence $s_i\, h(m) \equiv ws + (s_i + w_i)r \pmod q$;
(5) The Lee-Chang group signature on message m is given by (r, s, r_i, s_i).

A similar attack can also be mounted against the Tseng-Jan scheme but there is a simpler (universal) forgery. In the Tseng-Jan scheme, a group signature (r, s, A, B, C, D, E) must satisfy (see Eq. (2))

$$\alpha_i{}^{h(m)} \equiv DH_i{}^r r^s \pmod p$$

with $\alpha_i = D^B y_T{}^C E \bmod p$ and $DH_i = \alpha_i A \bmod p$. We first note that the value of α_i can be freely chosen if $E = \alpha_i (D^B y_T{}^C)^{-1} \bmod p$; likewise, the value of DH_i can be freely chosen if $A = DH_i \alpha_i{}^{-1} \bmod p$. Any adversary can thus computes a valid Tseng-Jan group signature as follows:

(1) Randomly choose r, ω, and compute $\alpha_i = r^\omega \bmod p$;
(2) Randomly choose B, C, D, and compute $E = \alpha_i (D^B y_T{}^C)^{-1} \bmod p$;
(3) Randomly choose σ and compute $DH_i = r^\sigma \bmod p$;
(4) Compute $A = DH_i \alpha_i{}^{-1} \bmod p$;
(5) Solve for s the congruence $\omega\, h(m) \equiv \sigma r + s \pmod q$;
(6) The Tseng-Jan group signature on message m is given by (r, s, A, B, C, D, E).

References

1. David Chaum and Eugène van Heijst. Group signatures. In *Advances in Cryptology — EUROCRYPT '91*, LNCS 547, pp. 257–265. Springer-Verlag, 1991.
2. Wei-Bin Lee and Chin-Chen Chang. Efficient group signature scheme based on the discrete logarithm. *IEE Proc. Comput. Digit. Tech.*, 145(1):15–18, 1998.
3. Anna Lysyanskaya and Zulfikar Ramzan. Group blind signatures: A scalable solution to electronic cash. In *Financial Cryptography (FC '98)*, LNCS 1465, pp. 184–197. Springer-Verlag, 1998.
4. Kaisa Nyberg and Rainer A. Rueppel. Message recovery for signature schemes based on the discrete logarithm problem. In *Advances in Cryptology — EUROCRYPT '94*, LNCS 950, pp. 182–193. Springer-Verlag, 1995.
5. Yuh-Min Tseng and Jinn-Ke Jan. Improved group signature scheme based on discrete logarithm problem. *Electronics Letters*, 35(1):37–38, 1999.

Security Properties of Software Components

Khaled Khan, Jun Han, and Yuliang Zheng

Peninsula School of Computing and Information Technology.
Monash University
McMahons Road, Frankston
VIC 3199 Australia.
Fax: +61-3-99044124
{khaled,jhan,yuliang}@mars.pscit.monash.edu.au

Abstract. This paper classifies security properties of software components into two broad categories: (1) non-functional security (NFS) properties, and (2) properties as security function (SF). Non-functional security properties are codified and embedded with the component functionality, whereas, properties as security functions are employed as external protection to the component. In most cases, users may add additional external protection to the binary form of the component. This classification could be used to determine how much the overall security of the component is dependent on the non-functional security properties of the component and to what extent the additional external protections are required in order to use the component in their specific application environment.

Keywords. Software component, Security properties, Component functionality

1 Motivation

Software component security is concerned with the protection of the resources, data, and computational properties of individual components and the enclosing system. The security of software component is very much dependent on the role that the component plays in a specific application environment. Software components need to be customised with the application environment where it is deployed. In most cases, third-party software components need external protection employed by the users to meet their specific application requirements. Due to the binary representation of the component, software composers are not able to modify the security properties that are embedded at the implementation level. We believe that overall security properties of software components should be classified into two categories, and this information should be available for users' inspection before a candidate component is selected.

There is a growing concern on the issue of characterising the component security properties in recent days as expressed in [2], [5], [6], [7] ,[8], [9]. The purpose of this paper is to identify various sources of security features that are codified with the component functionality. Most of the research conducted in recent days

M. Mambo, Y. Zheng (Eds.): ISW'99, LNCS 1729, pp. 52–56, 1999.
© Springer-Verlag Berlin Heidelberg 1999

related to component security are involved in detecting and removing the security flaws of components. We believe that security of component should be considered differently than the case of application systems because of the distributed usage of the component. In this paper, we are not going to propose any new security technique or assessment, rather we define different types of security properties that are placed various ways enforcing certain security objective of the component.

2   Classes of Component Security Properties

In this section, we define two distinguished classes of security properties related to software components such as *properties as security function (SF)* and *non-functional security (NFS)* properties.

Properties as security functions (SF) are those that provide secure encasing to the components, protecting the component from any security threat from the outside of its boundary. External security features can be added to components as functions. We call them *security functions (SF)*. On the other hand, *non-functional security (NFS)* properties are inherited in the internal implementation of the component, that is, some security features are codified with the functionality that the component provides. We term these features as *non-functional security properties (NFS)*. Usually, NFS properties are some security enforcement mechanisms that are already embedded in various forms with the functionality of the component at the implementation level. NFS properties are, in fact, the implementation of highly abstract security objectives that are intended to counter certain security threats. NFS properties are attached with various aspects of the component functionality in different layers of implementation, each representing a specific level of abstraction to achieve certain security objective. Whereas, SFs are considered some kind of external protection mechanisms of components which are not directly related to the component functionality. SFs can be designed and added to the existing component. For example, a wrapper can be written to protect the component from a potential security threat. The authentication of component origin and identity is considered as a SF. This type of SF protects the enclosing system from being assembled with an unauthorised component. Particularly in a dynamic assembly scenario, a target component may be located in a remote server whose identities may or may not be authenticated.

NFS properties are embedded with the component functionality to prevent the threat of usage violation. A component may employ certain NFS properties to guard its sensitive data and functionality from being violated by other unauthorised entities. The NFS properties embedded with the component's functionality may have substantial impact on the entire security mechanism of the composed system. It is not enough to make a system secure by adding SF such as encryption technology or secure protocols for data transmission to the component if the implementation of the component has security flaws. A great effort in designing SFs such as a strong authentication and access control mechanism can be easily ruined by a single security flaw embedded with the functionality [3]. If there is a weak NFS property in a sensitive operation, the rest of the strong SFs may not help much to protect the component. Authorisation mechanisms such as access control cannot always be guaranteed in a single access control mechanism, instead further protection may be required in terms of NFS properties to protect the functionality and the internal data of the component. In the next section, we will see how these two types of security properties enforce certain security objectives in various ways. To illustrate these two phenomena of component security properties we analyse a simple example in the following section.

3   An Example

In our example, *a medical system used by the medical practitioners is dynamically assembled with a software component time to time as needed. The component caters the diagnosis reports on patients to the legitimate users. ONLY selected users have access to the reports of his/her own patients although they may have access to the other services provided by the component.*

The service of the component in which we are interested in is a set of operations supported by an interface structure. The interface name and its associated attributes are visible to the users, but the operations are not. The various layers of SF and NFS properties identified in the example are illustrated in Fig. 1. We show in this figure how different security properties enforce certain security objectives by analysing the execution behaviour of the cited example. The circles denote various layers of security properties both as SF and NFS, and the arrows show the sequence of the execution behaviour of the component.

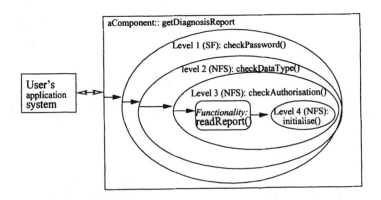

Fig. 1. SF and NFS Properties at Various levels

3.1   Identifying Security Properties

In the top level of the layer, we can find a SF concerning with the authentication and access control of the user to the component. The function is implemented by the operation *checkPassword()*, and used to protect the component from being accessed by any unauthorised entities. The function allows the user to assemble the component with the application system. It is not related to the functionality of the user system or the component, hence labeled as SF. Second level of the security is provided in terms of a NFS property implemented by the operation *checkDataType()*. This operation is closely related with the component functionality since it computes the output based on the input received from the user system. For a smooth processing, this NFS property is used to ensure that all input data values must be in correct type and range, but it has also a security side-effect that was not designed intentionally for security purposes. It is often reported that incompatibilities of the input values may lead to undesirable security problems of the component. Next level security in the hierarchy is provided by the operation *checkAuthorisation()* just before the main function *readReport()* of the component is executed. This operation does not employ any authentication or password to verify the authorisation, rather a tuple associated with the *patientID* in the database file verifies this. If the patient tuple in the database entry does not contain the *userID* value, the system would not authorise the user to read the data associated with the *patientID*. Finally, the security property supported by the operation *initialiseAllVariables()* in the layer 4 is a NFS that simply initialises all internal variables to enforce certain security objective, because users may leave a trail of *data shadow* with the contents of the data they manipulated [4].

3.2   Observations

We observe following characteristics from this simple example.

- some operations are implemented as underlying logical framework into the component functionality, and act as security side effects
- security properties propagate from high level to the low level security properties
- users interact directly with the SF
- users do not have direct interaction with the NFS properties
- users are unaware of the presence of the NFS properties unless any of their actions violates the NFS properties
- SF property acts as an external protection to the component
- NFS properties protect the internal data and operations as well as the functionality of the component.
- some of the NFS properties were not intentionally programmed for security purposes.

However, it is quite possible to uncover more NSF properties from this example if we analyse the implementation details of the operations, like call sequences, parameter passing, file permissions, and so on.

4   Summary and Further Work

In addition to the characterisation, all these extracted SF and NFS properties could be further assessed in terms of their strength and weakness. The presence of several SF and NFS properties in a functionality do not dictate that the function is secure, rather their relative strength and implementation mechanisms must be taken into active consideration. The characterisation and assessment of SF and NFS properties can be augmented with the *functional and assurance require-ments* defined in the *Common Criteria for the Information Technology Security Evaluation, version 2.0 ISO/IEC 15408* [1]. If we can manage to characterise the SF and NFS properties, and use them as component label with proper sealing technique using a digital signature, it may increase the user confidence on the component technology. The security characteristics on the label can also guide the user to decide whether a component would be able to meet the specific se-curity requirements of their application or not. The NFS information published on the component label would assist the user to design and add additional se-curity protection to the component to meet their specific security requirements. Simply adding security functions onto an existing component may not be as effective as if they were embedded with the component functionality. It is the internal computing properties that determine the ultimate behaviour of the com-ponent. We are currently exploring the possibility of applying Common Criteria requirements to software components to characterise the security properties of components. Such an effort may lead to defining a formal characterisation model that would help us to specify properly all SF and NFS properties of software components. Our research objective is to formalise a concrete characterisation framework, which could be used as the basis of software component certification.

References

[1] ISO-15408. Common Criteria for information technology security evaluation version 2.0. Standard, ISO/IEC 15408 NIST, USA, June 1999. http://csrc.nist.gov/cc/.

[2] U. Linquist and E. Jonsson. A map of security risks associated with using COTS. *IEEE Computer*, pages 60–66, June 1998.

[3] G. McGraw. Software assurance for security. *IEEE Computer*, pages 103–105, April 1999.

[4] A. Rubin and D. Geer. A survey of Web security. *IEEE Computer*, pages 34–41, September 1998.

[5] C. Szyperski. *Component Software -Beyond Object-Oriented Programming*. Addison-Wesley, 1998.

[6] N. Talbert. The cost of COTS. *IEEE Computer*, pages 46–52, June 1998.

[7] C. Thomson, editor. *Workshop Reports. 1998 Workshop on Compositional Software Architectures*, Monterey, USA, January 1998.
http://www.objs.com/workshops/ws9801/report.html.

[8] J. Voas. Certifying off-the-shelf software components. *IEEE Computer*, pages 53–59, June 1998.

[9] J. Voas. The challenges of using COTS software in component-based development. *IEEE Computer*, pages 44–45, June 1998.

Methods for Protecting a Mobile Agent's Route

Dirk Westhoff[1], Markus Schneider[2], Claus Unger[1], and Firoz Kaderali[2]

[1] FernUniversität Hagen, D-58084,
Fachgebiet Praktische Informatik II
{dirk.westhoff, claus.unger}@fernuni-hagen.de
[2] Fachgebiet Kommunikationssysteme
{mark.schneider, firoz.kaderali}@fernuni-hagen.de

Abstract. In the world of mobile agents, security aspects are extensively being discussed, with strong emphasis on how agents can be protected against malicious hosts and vice versa. This paper discusses methods for protecting an agent's route information from being misused by sites en route interested in gaining insight into the profile of the agent's owner or in obstructing the owner's original goal. Our methods provide visited sites with just a minimum of route information, but on the other hand allow sites to detect modifying attacks of preceding sites. Though, under noncolluding attacks, all methodes presented provide a similar level of protection, they differ w.r.t. performance and the points of time when an attack can be detected.

1   Introduction

Mobile agents are becoming more and more important for Internet based electronic markets. In many scenarios, mobile agents represent customers, salesmen or mediators for information, goods and services [1]. Agents are autonomous programs, which, following a route, migrate through a network of sites to accomplish tasks or take orders on behalf of their owners. Without any protection scheme, a visited site may spy out an agent's data and thus collect information about the agent's owner, e.g. its services, customers, service strategies, collected data, etc. To avoid such a situation, the amount of data accessible for a visited site has to be restricted as much as possible [2].

In this paper, we concentrate on an agent's route information, i.e. the address list of sites to be visited during a trip. The owner provides its agent with an initial route. On its travel, the agent works through the route stage by stage. Protecting the route guarantees that none of the visited stations can manipulate the route in a malicious way or can get an overview of other sites which the agent's owner is contacting. Protecting an agent's route against manipulation ensures an agent's owner that its agent really visits all intended sites; a visited site, e.g., is not able to remove a competing following site from the route. Hiding an agent's route from other sites ensures all participating sites that, e.g., their communication and commerce relationships are hidden from other sites and that a site may not change its offer or quality of service according to such kinds of information. We

M. Mambo, Y. Zheng (Eds.): ISW'99, LNCS 1729, pp. 57–71, 1999.

extend our protection scenario by allowing legal route extensions. To provide a certain requested service, a site may need to co-operate with other sites, not forseen by the agent's owner. In such a situation a site may extend the original route in a controlled and safe way. The following example provides a rationale for our concept. Let an owner send an agent to visit a predefined sequence of shops to find out the best offer for an electronic device. Our methods guarantee that no shop may base its offer on its knowledge about which shops are included or not included in the agent's route. They further guarantee that no competing shop can be deleted from the route without being detected by other shops or the owner. If a shop has the device in stock with a Japanese description, it may request another site to offer the German description or to translate the Japanese description into German. It therefore provides the agent with its offer but at the same time extends the original route by inserting the translator site into the route. Route extension are restricted to detours, i.e. they always lead back to the extending site.

To repulse malicious programs like Trojan horses, the identity of an agent becomes known to all visited sites [3], anonymous agents [4] are not dealt with in this paper. In the following we use the concept and terminology of the ALOHA[1]-system [5].

2   Related Work

Methods that protect an agent against attacks can be categorized into those which prevent attacks and those which detect attacks.

Cryptographic traces [6] detect illegal modifications of an agent by a post-mortem analysis of data the agent collected during its journey. *State appraisal* [7] mechanisms protect an agent's dynamical components, especially its execution state, against modifications. When an agent reaches a new site, the appraisal function is evaluated passing as a parameter the agent's current state.

Tamper-proof-devices [8] are hardware based and therefore not suitable in open systems. Software based approaches include the *computation of encrypted functions* [9] and *code scrambling* [10]. Unfortunately, the first approach can only be applied to polynominal and rational functions. In [10] the code of an agent is re-arranged to disguise the agent's functionality. To some extent, our results are related to work done in the area of anonymity. There are several reasons for aiming at anonymity:

- to decrease a party's observability and to prevent an attacker from performing traffic analysis, e.g. determining the partners of a communication;
- to hide the sending party from the receiving party, even in bidirectional communications, by using untraceable return addresses.

---

[1] The ALOHA environment allows its users to easily define, send, receive and evaluate agents, and is implemented with the help of the AAPI-package. It is downloadable at http://www.informatik.fernuni-hagen.de/import/pi2/agents/aapi-homepage.html.

Existing work in the area of anonymity can be classified in various ways. There exist proposals for unconditionally untraceable message exchange, as presented in [19], as well as computationally untraceable proposals, e.g. Chaum's *Mixes* [20] or *Onion Routing* [11], [21], [22] as a special Mix architecture for dynamically building anonymous connections. Further proposals in the area of anonymity focus on special applications like electronic cash, e.g. in [23], [24]. These approaches use completely different techniques (e.g. blinding protocols) than those used in anonymous connections and concentrate on application specific security problems like preventing electronic coins from being spent several times. Onion Routing is proposed for mobile applications, e.g. location hiding of cellular phones [25], or private Web browsing [26].

In our approach, we hide the route information of a mobile agent in such a way that only the agent's home site knows all sites to be visited; to reduce the communication overhead, each site is informed about its predecessor and successor site. Hiding an agent's route entries against visited sites is closely related to techniques being used in anonymous connections, where network nodes are prevented from performing traffic analysis and revealing communication relationships. In both scenarios, the emphasis is on preventing certain passive attacks. In the area of mobile agents, active attacks have to be considered as well. The potential for active attacks stems from the fact that sites visited on an agent's journey may be adversaries or competitors. A visited site, e.g., may be interested in eliminating a competitor from the agent's route. Thus, to hide an agent's route, this paper concentrates on methods based on encapsulation or concatenation techniques and discusses their strenghts for preventing and detecting active attacks.

3   Framework

An agent is an autonomous program which acts on behalf of its owner. According to its route, it visits sites linked together via a communication network. An agent is created, sent, finally received and evaluated in its owner's *home context*. At a visited site, the agent is executed in a *working context*. To save costs, an agent usually does not return to its home context before it has worked off its route; thus during the agent's journey its home site has not to be connected to the communication network all the time. To forward an agent to its next site or to extend an agent's route, each visited site needs access to a certain part of the route. A site is not allowed to remove a not yet visited site from the initial route and thus, e.g., exclude sites from offering their services to the agent. When a visited site extends a route, all added sites must become aware of the fact that they have not been on the initial route, and thus may, e.g., restrict the agent's access rights and functions, e.g. for electronic cashing. When a site changes a route, the change must be uniquely be associated with the site. To avoid arousing suspicion, as soon as a site detects an attack against an agent, it has to send the agent back to its home context.

In the following, we present a concept which reduces the route information, that becomes visible to a visited site, to a minimum, and which protects the route against malicious changes. In addition the concept is flexible enough to handle route extensions during the agent's journey. This paper mainly focuses on security aspects and thereby demonstrates the basis for the actual implementation of the ALOHA system. In a second step we will concentrate on performance issues, e.g. by substituting the pure public-key encryption schemes as presented in this paper by combined symmetric/public-key encryption schemes. Such a substitution increases the performance of the methods presented but does not influence the security of route protection.

Because of network traffic, the concept of a trustworthy center to be visited between each two consecutive sites, is not discussed in this paper. Each site is only given access to the address of its predecessor and its successor site.[2]

Additionally to the route, the agent includes other components like *profile*, *binary code*, *mobile data* and a *trip marker* per journey. Agents have to be protected against passive, reading attacks, and active attacks that modify an agent's functionality or even fully destroy it. This paper concentrates on the route and its protection against collusion-free attacks.

4   Protecting the Initial Route

When an agent migrates from working context c_i to working context c_{i+1}, all its objects are encrypted and thus protected against passive attacks. Before starting an agent, (except in very special cases [9]) a working context has to decrypt parts of the agent and thus make the agent vulnerable against active or passive attacks. Thus the agent's route should be protected as much as possible.

An unprotected route $r = ip(c_1) \, || \, \ldots \, || \, ip(c_n)$ is a concatenated list of working context's Internet addresses $ip(c_i)$ that have to be visited during an agent's journey. To abort an agent's journey, each site has to know the Internet address $ip(h)$ of the home context h, which therefore is stored in plaintext separate from the protected route.

In the following subsections, we present several combinations of encryption and signature schemes and compare their properties. In all proposed methods, before starting an agent, the home context generates the protected route. It uses public-key encryption and applies concatenation or encapsulation techniques in such a way that every visited working context by decryption can only reveal those data (e.g. the successor's address) that are relevant for itself. Furthermore, the visited working context verifies whether the received data are compromised by means of a signature generated by the agent's home context. Before forwarding the agent to the following working context, the actual working context deletes route destined for itself.

---

[2] A site does not necessarily need to know the address of its predecessor; on the other hand, the Internet communication protocol itself automatically provides a receiver with the address of the sender.

In the area of public-key cryptography, the general problem of providing authentic keys is well-known. In this paper, we don't approach this problem but in the following, we assume that there exists a secure public-key environment, including the generation, certification and distribution of public keys; each working context is assumed to know the authentic public keys of all other working contexts.

4.1   Atomic Encryptions and Signatures

If the home context encrypts and signs an agent's route in an atomic way (see equation (1)) by using an public-key encryption method E and a signing method S, the actual working context merely can decrypt its successor's Internet address and check the route for modifications.[3]

In the following, we first present the structure of the protected route r generated by the home context as it is shown in equation (1). The symbol $\|$ denotes the concatenation of data. Having presented the protected route as a whole, we explain and justify the route's structure step by step.

$$
\begin{aligned}
r = \ & E_{e_1}\big[ip(c_2), S_h\big(ip(h), ip(c_1), ip(c_2), t\big)\big] \ \| \\
& \vdots \\
& E_{e_{n-1}}\big[ip(c_n), S_h\big(ip(c_{n-2}), ip(c_{n-1}), ip(c_n), t\big)\big] \ \| \\
& E_{e_n}\big[EoR, S_h\big(ip(c_{n-1}), ip(c_n), EoR, t\big)\big]
\end{aligned}
\tag{1}
$$

In this case the route contains the encrypted addresses, respectively their signatures, of all working contexts that have to be visited. Because of the minimal number of data to be encrypted, this variant is very cost effective.

A working context c_i decrypts the Internet address $ip(c_{i+1})$ of its successor c_{i+1} by using its private key d_i that corresponds to its public key e_i. Each site removes its decrypted address and signature from the agent's route. All the other route entries are hidden from the actual working context.

With the help of the digital signatures, active attacks can be detected. The signature of the successor's address $ip(c_{i+1})$ which is signed by the agent's home context h and presented to the actual working context c_i proves that the route entry for c_i has not been modified. Internet address $ip(c_i)$ and the trip marker t guarantee a visited working context that itself is part of the initial route. The trip marker t is unique for each journey. It uniquely identifies an agent's journey and prevents replay attacks. Otherwise, a malicious working context could replace the complete route of the actual agent with a copied route of an agent's earlier journey.

We illustrate the meaning of t by an example. Suppose, an owner regularly sends an agent to visit sites along a standard route. Without a trip marker t, an

---

[3] The ALOHA-system uses RSA [12] as public-key encryption E. The signature S is based on a combination of SHA [13] and RSA.

attacker could save a copy of this standard route and, when the owner changes the standard route, replace the new route by the old one. In a scenario without unique trip markers, a visited site is not able to detect whether the actual route has been replaced by a copy of a valid old route created – and also signed – by the same home context. Under this attack, a verification of the signature in the copied route would prove that the compromised route really stems from the actual home context. The signature would also make the visited site believe that it is contained in the actual route and convince him to offer the requested service. Without trip markers, replacing a route with a valid copy can only be detected when the agent arrives at his home context, delivering data from unexpected sites. It would now be the home context's duty to convince a visited site that the route was forged. Especially in an electronic commerce environment, such attacks have to be prevented by all means. A unique trip marker t, e.g. the agent's start time, allows every visited site to verify the originality of the agent's route. When a visited site receives from the same home site agents with identical trip markers, it immediately detects the replay attack.

The importance of the predecessor's address $ip(c_{i-1})$ in each signature becomes obvious in the following scenarios. With the help of the EoR entry, working context c_n realizes that itself is the final entry of the agent's route. Via EoR and t in the signature, working context c_n is able to check whether these data have been generated by h and whether itself was really included in the initial route.

Signatures must be encrypted by the home context as well, otherwise, if the agent carried the signatures in plaintext, an attacker who knows t would be able to reconstruct the complete route by arranging and testing combinations of known addresses. Such an attack is feasible if the number of relevant sites is small.

In the following we discuss the method with regard to active attacks which modify the functionality of an agent. If a malicious working context c_i removes entries from the agent's route [see fig. 1], this change will be detected by its new successor c_j when it checks the signature. In such a case, c_j sends the agent back to its home context. The predecessor's address has to be included in the signature. Otherwise, if a malicious context c_i could predict another valid working context c_j, $i < j \leq n$, and its position in the encrypted route, it could skip over $c_{i+1}, \ldots, c_{j-1}$, just by removing these entries from the agent's route. It could directly send the agent, together with its modified route, to context c_j, without giving c_j a possibility to detect this attack.

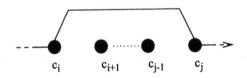

Fig. 1. Remove of addresses.

Of course, the chance for such an attack depends on the number of potential working contexts. It may be easier for a malicious working context to predict a working context in an agent system with few providers of comparable services than in an agent system with a lot of providers offering the same services.

If a malicious context tries to insert m ciphertexts of new Internet addresses $c_1^M, \ldots, c_m^M$ after removing $j - i$ entries from the route [see fig. 2], then working context c_1^M will detect this modification by signature check. If context c_1^M follows the rules of the agent system, it stops the agent's journey and sends it back to its home context. Supposing the malicious working context forms a collusion with all added contexts $c_1^M, \ldots, c_m^M$, probably none of these contexts will abort the agent's journey. In any case such an attack will be detected when context c_j on the initial route checks the signature.

Instead of removing or exchanging subsequent entries from the agent's route a malicious context may be interested in removing or exchanging later entries. In such a situation the question arises when such an attack can be detected at the earliest. The later an attack is detected the more contexts may become suspicious to be a dishonest context from the home context's point of view.

Consider the following scenario: the malicious working context c_i removes entry $E_{e_j}[ip(c_{j+1}), S_h(ip(c_{j-1}), ip(c_j), ip(c_{j+1}), t)]$ destined for c_j and the following $m - 1$ entries with $0 < m < n - j$ from the route.

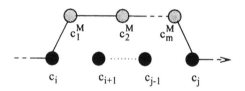

Fig. 2. Exchange of addresses.

At the beginning, this kind of modification does not attract any attention. Only working context c_j can detect this attack, because its route entry can only be decrypted by c_{j+m}. After c_j has decrypted the ciphertext, it recognizes that the plaintext does not contain a valid Internet address. Thus, c_j aborts the agent's journey and sends it back to its home context.

If c_i removes the last $m = n - j$ entries from the agent's route, then the working context c_j, after decoding the last entry, gets the Internet address $ip(c_{j+1})$ instead of EoR. Again c_j is the first context which is able to recognize the attack.

If c_i exchanges existing entries by new entries, e.g. by exchanging the initial route's $ip(c_{j+1})$ by $ip(c_1^M)$ and further entries, again c_j can detect this attack at the earliest.

Summing it up: atomic encryption and signature sufficiently protects the agent against all presented attacks. Unfortunately several attacks can only be detected by contexts quite far away from the attacking context and the attacking context itself cannot uniquely be identified. In such a situation, the agent can perform work at sites not forseen in the initial route before the attack is detected.

4.2 Atomic Encryptions, Nested Signatures

This approach (see (2)) means higher computational complexity than the first one.

$$
\begin{aligned}
r = {} & E_{e_1}\big[ip(c_2), S_h\big(ip(h), ip(c_1), ip(c_2), E_{e_2}[\ldots], \ldots, E_{e_{n-1}}[\ldots], E_{e_n}[\ldots], t\big)\big] \; \| \\
& E_{e_2}\big[ip(c_3), S_h\big(ip(c_1), ip(c_2), ip(c_3), E_{e_3}[\ldots], \ldots, E_{e_{n-1}}[\ldots], E_{e_n}[\ldots], t\big)\big] \; \| \\
& \qquad\qquad\qquad\qquad\qquad\qquad\qquad\qquad\qquad\qquad\qquad\vdots \\
& E_{e_{n-1}}\big[ip(c_n), S_h\big(ip(c_{n-2}), ip(c_{n-1}), ip(c_n), E_{e_n}[\ldots], t\big)\big] \; \| \\
& E_{e_n}\big[EoR, S_h\big(ip(c_{n-1}), ip(c_n), EoR, t\big)\big]
\end{aligned}
\tag{2}
$$

For the actual context c_i, all secret addresses and the signature destined for c_i as well are encrypted. The signature contains the address of the actual working context, the predecessor's and the successor's address, the trip marker t, as well as the ciphertexts to be used by later sites, including the ciphertext EoR.

A visited working context decrypts its successor's address and its specific signature. Solely, this context itself can with its private key decrypt these data. Via the signature the context checks whether the route's entries have been modified and with the signature that depends on its own Internet address whether itself is part of the initial route.

If the signature verification is successful, the context removes the encrypted successor's address and its signature from the agent's route and thus generates the route for the next migration of the agent. A successfully checked signature also guarantees the context that it belongs to the initial route of the agent.

The predecessor's address in the signature allows to detect attacks as presented in fig. 1. Otherwise, if a malicious context c_i correctly predicts a later entry c_j of the agent's route, it could eliminate all entries before entry c_j.

Like in our first approach, an attack according to fig. 2 can be detected by context c_1^M.

Both approaches discussed so far equally protect the agent's route against the attacks described, but they differ in the case when a malicious context c_i removes or exchanges entries not adjacent to its own entry. With the second approach, all such attacks can always be directly detected in c_{i+1} by signature verification. Thus, some of the weaknesses of the first approach can be avoided at higher computational complexity.

4.3 Nested Encryptions, Atomic Signatures

With this approach the actual working context c_i receives a sole ciphertext from its predecessor's context. By decrypting this ciphertext the actual working context gets its successor's address, the signature destined to c_i by the home context and the ciphertext for the successor's context (see (3)). After the working context has checked and evaluated the two first components, it removes these data from the agent's route and sends the remaining ciphertext to its successor.

$$r = E_{e_1}\left[ip(c_2), S_h\big(ip(c_1), ip(c_2), t\big), E_{e_2}\left[ip(c_3), \ldots, \right.\right.$$

$$\left.\left. E_{e_{n-1}}\left[ip(c_n), S_h\big(ip(c_{n-1}), ip(c_n), t\big), E_{e_n}\big[EoR, S_h\big(ip(c_n), EoR, t\big)\big]\right] \ldots \right]\right] \quad (3)$$

In contrast to the two preceeding approaches, in this approach a signature does not need to contain the predecessor's address. From the description of the previous approaches, it is clear why the actual address, the predecessor's address and the trip marker t have to be included into the signature. Because of the nested encryption, an attack from fig. 1 can be detected even without including the predecessor into the signature. If a malicious context c_i attempts to delete all sites between itself and c_j, it is not able to offer c_j valid information according to the described protocol. The same is true if it attempts to substitute these sites by sites of its own choice [see fig. 2]. Both attacks are detected by the next working context that is visisted by the agent.

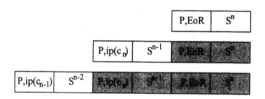

Fig. 3. Route after its decryption at block layer.

If a context modifies entries not adjacent to its own entry in many cases such an attack cannot immediately be detected but may destroy route entries for later sites. To explain this it is useful to have a look at the underlying data structures: When an encrypted message is changed, the corresponding decrypted message is completely changed. RSA [12] is a block based encryption scheme, i.e. changing an encrypted block only affects the corresponding decrypted block. A route entry with successor address and signature covers two blocks [see fig. 3], where S^i denotes the signature destined for c_i and P denotes a padding pattern. The grayed cells denote encrypted information, the white ones plaintext. The information of the top line is destined for c_n, the information of the next line for c_{n-1}, etc. If a malicious context c_i modifies the blocks at the end of a line, only c_n could detect such an attack. If it deletes these blocks, c_{n-1} would detect such an attack because it expects EoR. In both cases such an attack is detected very late. In general one can observe an analogous property concerning the earliest possibility of detection if the attack is on a preceding block.

Summing it up: even the higher computational complexity of this approach does not guarantee that attacks are detected as early as possible, though the fact that the predecessor's address is not needed for detecting some attacks may mean an advantage.

4.4   Nested Encryptions and Signatures

This approach includes the advantages of both previous approaches, in exchange for higher computational costs (see (4)).

$$
r = E_{e_1}\Bigg[ip(c_2), S_h\big(ip(c_1), ip(c_2), t, E_{e_2}[\ldots]\big), E_{e_2}\Bigg[\ldots, \tag{4}
$$

$$
E_{e_{n-1}}\Big[ip(c_n), S_h\big(ip(c_{n-1}), ip(c_n), t, E_{e_n}[\ldots]\big), E_{e_n}\big[EoR, S_h\big(ip(c_n), EoR, t\big)\big]\Big]\ldots\Bigg]\Bigg]
$$

Working context c_i decrypts the address of its successor, a signature and the ciphertext to be transmitted to the next context. Before sending off the agent, all plaintexts are removed from the route. The signature contains the actual working context's address, its sucessor's address, the trip marker t and the complete rest of the route. To avert attacks, the predecessor's address is not needed. Analogously to the consideration we made for the other approaches, all attacks can be detected as early as possible.

4.5   Summary

All presented approaches protect the agent's route against the described attacks. Only variants (2) and (4) with nested signatures guarantee that an attack is detected as early as possible, where variant (2) is more efficient with regard to its computational complexity.

Because of its lower costs, the first approach (1) may be used for short routes or less sensitive services. Although not discussed in this paper, variant (3) and (4) with nested encryptions are suitable to protect the agent's route against attacks performed by collusions of malicious contexts.

With all alternatives, if the home context receives its agent from context c_n without an error message, it can be sure that its agent visited all contexts of the initial route and as much adresses as possible are kept secret.

In the following we will examine cases where routes are legally extended during an agent's journey.

5 Extending the Initial Route

If for providing its service a context c_i on the agent's initial route needs the cooperation with other contexts $c_1^X, \ldots, c_m^X$, then working context c_i should be allowed to extend the initial route in a legal way. Of course c_i should not be allowed to delete unvisited entries from the initial route.

To protect the home context's interests, the new entries do not include any confidential information.[4] On the other hand, c_i may be interested in protecting its new entries. Let $c_1^X, \ldots, c_m^X$ be these new entries. Like a home context, c_i encrypts the new route extension and includes it in accordance to the choosen approach as a prefix to the initial route (see (5) with variant 2 (atomic encryption and nested signatures) as example).

$$
r^X =
$$

$$
E_{e_{X1}}\left[ip(c_2^X), S_i\big(ip(c_i), ip(c_1^X), ip(c_2^X), E_{e_{X2}}[..], .., E_{e_{Xm}}[..], E_{e_i}[..], .., E_{e_n}[..], t\big)\right] \;\|
$$
$$
E_{e_{X2}}\left[ip(c_3^X), S_i\big(ip(c_1^X), ip(c_2^X), ip(c_3^X), E_{e_{X3}}[..], .., E_{e_{Xm}}[..], E_{e_i}[..], .., E_{e_n}[..], t\big)\right] \;\|
$$
$$
\vdots
$$
$$
E_{e_{Xm-1}}\left[ip(c_m^X), S_i\big(ip(c_{m-2}^X), ip(c_{m-1}^X), ip(c_m^X), E_{e_{Xm}}[..], E_{e_i}[..], .., E_{e_n}[..], t\big)\right] \;\|
$$
$$
E_{e_{Xm}}\left[EoX, S_i\big(ip(c_{m-1}^X), ip(c_m^X), EoX, E_{e_i}[..], .., E_{e_n}[..], t\big)\right] \;\|
$$
$$
E_{e_i}\left[ip(c_{i+1}), S_q\big(ip(c_{i-1}), ip(c_i), ip(c_{i+1}), E_{e_{i+1}}[..], .., E_{e_n}[..], t\big)\right] \;\|
$$
$$
\vdots
$$
$$
E_{e_{n-1}}\left[ip(c_n), S_q\big(ip(c_{n-2}), ip(c_{n-1}), ip(c_n), E_{e_n}[..], t\big)\right] \;\|
$$
$$
E_{e_n}\left[EoR, S_q\big(ip(c_{n-1}), ip(c_n), EoR, t\big)\right] \quad (5)
$$

If c_i extends a route it must not delete the ciphertext representing its sucessor and signature from the initial route. Further on, c_i includes, with its signature, the information that itself has changed the initial route.

Having visited all sites $c_1^X, \ldots, c_m^X$ of a route extension, the presented concept provides that the agent returns to c_i [see fig. 4].

When the agent returns to c_i, the context can decrypt its original part of the initial route, check it, and send the agent to the next site on the initial route if the check is successful.

The route extension follows one of the four protection variants, and thus inherits their protection properties.

All signatures in the route extension stem from c_i; the route extension ends with an EoX entry. If a context receives such an EoX entry, it sends the agent back to c_i. To detect attacks as early as possible, the remaining ciphertexts of the initial route are included into c_i's signature. Thus, when the agent returns from c_m^X to c_i, it can be sure that all sites of the route extension have been visited.

---

[4] The new contexts get just informed that c_i is on the initial route.

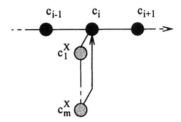

Fig. 4. Extending a route.

This concept of route extensions can be extended: each context of a route extension may be allowed to extend the route extension itself as well. In all these cases our scheme guarantees that either all sites in the initial route as well as all route extensions are visited, or the journey is explicitly aborted because of an attack or a not accessible site.

6 Unreachable Working Contexts

In many systems, the gain of technical data protection conforms with a loss of functionality and flexibility. If a working context c_{i+1} is unavailable, because it terminated regularly/irregularly or the network connection was interrupted, then context c_i, that can decrypt only the Internet address $ip(c_{i+1})$, may not just skip over context c_{i+1}. In such a case, context c_i has to choose one of the following strategies:

- As long as c_{i+1} is unreachable, the agent waits in c_i.
- The agent aborts its journey and migrates back to its home context.
- Having tried to reach c_{i+1} for a certain time, the agent migrates back to its home context.

When an agent aborts its journey, the reason should become obvious to its home context. When the probability to reach a context falls below a certain level, the route protection mechanism should be changed to become more flexible, though from the data protection point of view this may be disadvantageous.

7 Conclusions

Although all the presented variants are able to resist any of the described attacks, only the two variants with nested signatures detect attacks as early as possible. From these both variants the one using an atomic encryption requires less computation time. Even the variant with atomic encryptions and atomic signatures can successfully be used under certain conditions.

Each variant allows legal route extensions. If a context extends an agent's initial route, having visited all sites of the route extension the agent returns to the context that has extended the route, and proceeds on its initial route. Following this pattern, one can develop several levels of route extension and use the described methods for protecting the agent's route against attacks. All the presented variants are realized in the ALOHA-system.

8 Future Work

To precisely identify that context which ran the attack, more in-depth research is needed.

While this paper mainly focuses on security properties of the proposed methods, a detailed performance analysis of these methods has to be done, taking into account various parameters like average route length, public key size, combined symmetric/public key encryption. In the present implementation state of our system, the total processing time for the route (decryption, signature verification) with 5 entries applying exclusively public-key cryptography with 1024 bit key size is about 5 times $1.42 - 1.87$ *sec* for the approaches (1) and (2) (Pentium AMD K6II 300MHz, 96MByte EDO-RAM, Windows 95, JDK 1.1.7B). In cases (3) and (4), the processing time (decryption, signature verification) decreases from 12.14 *sec* in the first working context to 1.48 *sec* for the fifth context on the journey.

We are also extending our route protection mechanisms to other agent components. Mobile data [14] can be handled similar to routes: at least parts of them must only become accessible to specific contexts. In contrast, the agent's binary code has fully to be decrypted at each site. By using checksums and a kind of third party protocol [15],[16], attacks can be detected and afterwards malicious contexts can be forced to forward correct binary code. Nevertheless such a protocol can not verify if a working context really started an agent. Maybe detection objects [17] can ensure such a property.

In addition we are working on methods protecting agents from colluding attacks. In [18], we consider a more restrictive trust model compared with the fourth approach in this paper. Dealing with a collusion framework it becomes relevant, e.g. if different sites to be visited on an agent's journey are provided by the same entity. In general, the fact that different sites are offered by the same provider (e.g. two shops belonging to the same company) is not obvious for a home context. Thus, there arise constellations in which further security analysis of route protection methods are necessary.

References

1. F. Mattern: 'Mobile Agenten', it + ti - Oldenbourg Verlag, 1998, 4, pp.12-17.
2. W. Ernestus, D. Ermer, M. Hube, M. Köhntopp, M. Knorr, G. Quiring-Kock, U. Schläger, G. Schulz: 'Datenschutzfreundliche Technologien', DuD 21, 1997, 12, pp.709-715.
3. S. Berkovits, J. Guttman, V. Swarup: 'Authentication for Mobile Agents', in 'Mobile Agents and Security', Proceedings, Springer Verlag, LNCS 1419, 1998, pp.114-136.

4. D. Chess: 'Security Issues in Mobile Code Systems', in 'Mobile Agents and Security', Proceedings, Springer Verlag, LNCS 1419, 1998, pp.1-14.

5. D. Westhoff: 'AAPI: an Agent Application Programming Interface', Informatikbericht 247-12/1998, FernUniversität Gesamthochschule in Hagen 1998.

6. G. Vigna: 'Cryptographic Traces for Mobile Agents', in 'Mobile Agents and Security', Proceedings, Springer Verlag, LNCS 1419, 1998, pp.138-153.

7. W.M. Farmer, J. Guttman, V. Swarup: 'Security for Mobile Agents: Authentication and State Appraisal', in 'Proc. of the 4th European Symp. on Research in Computer Security', Springer Verlag, LNCS 1146, 1996, pp.118-130.

8. U.G. Wilhelm: 'Cryptographically protected Objects'. Technical report, Ecole Polytechnique Federale de Lausanne, Switzerland, 1997.

9. T. Sander, C. Tschudin: 'Protecting Mobile Agents Against Malicious Hosts', in 'Mobile Agents and Security', Proceedings, Springer Verlag, LNCS 1419, 1998, pp.44-60.

10. F. Hohl: 'Time Limited Blackbox Security: Protecting Mobile Agents from Malicious Hosts', in 'Mobile Agents and Security', Proceedings, Springer Verlag, LNCS 1419, 1998, pp.92-113.

11. M.G. Reed, P.F. Syverson, D.M. Goldschlag : 'Anonymous Connections and Onion Routing', in 'IEEE Journal on Selected Areas in Communication - Special Issue on Copyright and Privacy Protection', Vol. 16, No. 4, 1998, pp.482-494.

12. R. Rivest, A. Shamir, L. Adleman: 'A Method for Obtaining Digital Signatures and Public-Key Cryptosystems' in 'Communication of ACM', Volume 21, Number 2, February 1978, pp.120-126.

13. A. Menezes, P. van Oorschot, S. Vanstone: Handbook of Applied Cryptography, CRC Press, 1996.

14. G. Karjoth, N. Asokan, C. Güclü: 'Protecting the Computation Results of Free-Roaming Agents', in 'Mobile Agents', Proceedings, Springer Verlag, LNCS 1477, 1998, pp.195-207.

15. N. Asokan, V. Shoup, M. Waidner: 'Asynchronous Protocols for Optimistical Fair Exchange', 1998 IEEE Symposium on Research in Security and Privacy, IEEE Computer Society Press, Los Alamitos 1998, pp.86-99.

16. Y. Han: 'Investigation of Non-repudiation Protocols', in 'Information Security and Privacy', Proceedings of ACISP'96, Springer Verlag, LNCS 1172, 1996, pp.38-47.

17. C. Meadows: 'Detecting Attacks on Mobile Agents', Center for High Assurance Computing Systems, Naval Research Laboratory, Washington DC, 1997.

18. D. Westhoff, M. Schneider, C. Unger, F. Kaderali: 'Protecting a Mobile Agents Route against Collusions', Sixth Annual Workshop on Selected Areas in Cryptography 99, August 99, Queens University, Kingston, Canada, Proceedings, (to appear at Springer Verlag, LNCS Series).

19. D. Chaum: 'The Dining Cryptographers Problem: Unconditional Sender and Recipient Untraceability', Journal of Cryptology, 1988, 1, pp.65-75.

20. D. Chaum: 'Untraceable Electronic Mail, Return Addresses and Digital Pseudonyms', Communications of the ACM, Vol. 24, No. 2, 1981, pp.84-88.

21. D. Goldschlag, M. Reed, P. Syverson: 'Hiding Routing Information', Information Hiding, Proceedings, Springer Verlag, LNCS 1174, 1996, pp.137-150.

22. D. Goldschlag, M. Reed, P. Syverson: 'Onion Routing for Anonymous and Private Internet Connections', Communications of the ACM, Vol. 42, No. 2, 1999, pp.39-41.

23. D. Chaum: 'Privacy Protected Payments – Unconditional Payer and/or Payee Untraceability', Smartcard 2000, Elsevier, 1989, pp.69-93.

24. S. Brands: 'Untraceable Off-line Cash in Wallet with Observers', Crypto 93, Proceedings, Springer Verlag, LNCS 773, 1993, pp.302-318.
25. M. Reed, P. Syverson, D. Goldschlag: 'Protocols Using Anonymous Connections: Mobile Applications', Security Protocols, Proceedings, Springer Verlag, LNCS 1361, 1998, pp.13-23.
26. P. Syverson, M. Reed, D. Goldschlag: 'Private Web Browsing', Journal of Computer Security, Vol. 5, No. 3, 1997, pp.237-248.

Non-interactive Cryptosystem for Entity Authentication

Hyung-Woo Lee[1], Jung-Eun Kim[2], and Tai-Yun Kim[2]

[1] Dept. of Information & Communication Engineering, Chonan University, 115, Anseo-dong, Chonan, Chungnam, 330-180, Korea.
Hwlee@infocom.chonan.ac.kr
http://infocom.chonan.ac.kr/~hwlee/index.html
[2] Dept. of Computer Science & Engineering, Korea University, 1, 5-ga, Anam-dong, Seongbuk-ku, Seoul, 136-701, Korea.
{jekim,tykim}@netlab.korea.ac.kr

Abstract. In case of mobile agent based computing system such as agent-based electronic payment and online electronic publishing of multimedia contents, both precise identification and secure authentication schemes are required for its security. The public-key cryptosystem and the digital signature scheme have been the foundation of overall secure systems. The requirement for providing agent based secure digital contents in electronic commerce is to implement the compatible secure entity authentication scheme. In this paper, existing discrete logarithm based Schnorr like authentication schemes are improved by the analysis of performance and security on the interactive protocols. And ElGamal type authentication schemes are also proposed. Then, they are enhanced with oblivious transfer based non-interactive public key cryptosystem for entity authentication. Proposed non-interactive protocols are applicable to the non-interactive zero knowledge proofs and they can provide compatible performance and safety in distributed commerce applications such as copyright protection system on multimedia contents.

1   Introduction

We must provide mutual authentication between mobile entities for providing secure services in electronic payment systems[1,2,3,4] and multimedia contents providing[28,29]. Public key based mutual authentication protocol can be applicable to the diverse applications in distributed commerce and access control primitives for the shared resources[5,6]. Existing authentication schemes[11,12] can be generalized to the interactive public key cryptosystem based on zero knowledge proofs[16]. However, we need to modify those interactive-type authentication schemes for optimizing the overall performance and for enhancing the communication bandwidth of entities such as digital contents provider in mobile

---

University Research Program supported by Ministry of Information & Communication in South Korea, 1999.

M. Mambo, Y. Zheng (Eds.): ISW'99, LNCS 1729, pp. 72-84, 1999.

electronic commerce applications[27]. With the growing of performance, the security of each entity must be strengthened to receive a transparent service. Therefore, we modify existing authentication schemes for enhancing the security and then change them into the non-interactive public key cryptosystem[19,20,22] with additional secure protocol for enhancing its performance and safety.

In this study, modified entity authentication methods are proposed based on the analysis of the existing schemes. Interactive authentication protocols are improved to the generalized schemes for enhancing the performance and security. And we apply these schemes to the *ElGamal*[14] like method. And we propose non-interactive authentication schemes connected with *oblivious transfer*[18] protocol, which are developed as a *non-interactive zero knowledge proofs*. Suggested non-interactive authentication protocols can be directly applicable to the certification of mobile entities and multimedia content providing system after the analysis of security and performance evaluations.

We now briefly describe the organization of this paper. We first present the existing authentication protocols and review both interactive and non-interactive proofs in section 2. We propose non-interactive public key cryptosystem for entity authentication in section 3 and then apply them to non-interactive zero knowledge proofs in section 4. In section 5, we analyze the safety of these schemes and evaluate their performances. We conclude this study with the future works in section 6.

2   Non-interactive Protocols

Interactive protocol is the process whereby one party is assured through acquisition of corroborative evidence of the identity of a second party involved in a protocol and that the second has actually participated[9,10]. In the case of honest parties A and B, A is able to successfully authenticate itself to B. And B can't reuse an identification exchange with A so as to successfully impersonate A to a third party C. The existing interactive protocols are based on the intractability of the discrete logarithm problem. The design allows pre-computation, reducing real-time computation for the claimant to one multiplication modulo a prime q. An important computational efficiency results from the use of a subgroup of order q of the multiplicative group of integers modulo p, where $q|(p-1)$. This also reduces the required number of transmitted bits. This protocol was designed to require only three passes, and a low communication bandwidth. The basic idea is that A (the *prover*) proves knowledge of a secret s to B (the *verifier*) without revealing his secret in a time-variant interactive manner depending on a *challenge* c in common client/server communication systems.

Okamoto[8] presents a *Schnorr*-type interactive authentication scheme and enhances the security of *Schnorr* authentication scheme. It is almost as secure as the discrete logarithm problem. A prover generates his public key $\{p,q,g_1,g_2,t,v\}$ and a secret key $\{s_1,s_2\}$. Using $\{s_1,s_2\}$ pairs, which are corresponded to the s in *Schnorr* scheme, public information $v \equiv g_1^{-s_1} g_2^{-s_2} \bmod p$ is computed. The operation steps of *Okamoto* are similar with those of *Schnorr*.

Existing non-interactive protocol[20] proposed the shared random string model between the prover and the verifier for *non-interactive zero knowledge proofs*. In this model, the prover and the verifier are given a shared string of bits that are guaranteed to be completely random. Then, the prover send a zero knowledge proof of a theorem as a single message to the verifier who can then check the proof without further interaction[21,22,23]. Such non-interactive zero knowledge proofs turned out to be an important cryptographic primitive in diverse applications.

And the concept of an *oblivious transfer* was introduced by *Rabin*[18]. As A send m bits in a message to B in 50% probability, A does not know which bits were sent. We can define *oblivious transfer* (*OT*) protocol as follow. A has two string s_0, s_1. A encrypts one of his strings and message m with B's public key P_B. And A sends this to B. B decrypts it and extracts one of the strings s_0 or s_1 using his private key. But A will not know which of the two B got. [19] developed this protocol into a simple and concrete non-interactive mode based on the *Diffie-Hellman* assumption[5].

In this study, we suggest the interactive authentication schemes and develop them as non-interactive manners. We aim at the enhancements of performance and safety in multimedia content providing system and electronic commerce applications on which secure authentication transactions are required.

3   Non-interactive Entity Authentication

If the existing interactive schemes are improved into one way mono-directional non-interactive protocols, its performance will be much enhanced in considering the bound of computation and its performance in secure objects together with its communication bandwidth.

So, we remove the bi-directional interaction in existing *Schnorr*-type authentication schemes. For doing it, we apply non-interactive oblivious transfer protocol. Using B's secret key (i, x_i), the verifier computes his public key β_i, β_{1-i}. After selecting two random number z_0, z_1, the prover computes τ_0, τ_1 such that $\tau_0 \equiv \beta_0^{z_0} \bmod p$, $\tau_1 \equiv \beta_1^{z_1} \bmod p$. And the prover generates $\sigma_0, \sigma_1 \in \{0,1\}^k (k = |p|)$ which satisfies both $\langle \tau_0, \sigma_0 \rangle = 0$ and $\langle \tau_1, \sigma_1 \rangle = 1$. Selected σ_0, σ_1 are the shared reference strings for non-interactive proofs between the prover and the verifier. By sharing these strings in advance, the prover does not need any *challenge* from the verifier. And the prover can generate his zero knowledge proof using these shared reference strings in equation (1).

$$y_i^e \equiv \begin{cases} y_i^0 \equiv (s_i \cdot \sigma_0 + w_i) \bmod q \\ y_i^1 \equiv (s_i + w_i \cdot \sigma_1) \bmod q \end{cases} (1 \le i \le k) \qquad (1)$$

We propose the non-interactive authentication scheme based on the oblivious transfer on the modified interactive protocol in Fig. 1.

Prover(A)	Verifier(B)

$s_i < p - 1 (1 \le i \le k)$, $s \equiv \displaystyle\sum_{i=1}^{k} s_i \bmod q$ $\qquad\qquad t \equiv g^s \bmod p$

$\qquad\qquad\qquad\qquad i \in \{0,1\}$, $x_i \in \{0,...,p-2\}$

$\qquad\qquad\qquad \beta_i \equiv g^{x_i} \bmod p$, $\beta_{1-i} \equiv c \cdot (g^{x_i})^{-1} \bmod p$

$z_0, z_1 \in \{0,...,p-2\}$

$\tau_0 \equiv \beta_0^{z_0} \bmod p$, $\tau_1 \equiv \beta_1^{z_1} \bmod p$

$\sigma_0, \sigma_1 \in \{0,1\}^k (k = |p|)$

$\langle \tau_0, \sigma_0 \rangle = 0, \langle \tau_1, \sigma_1 \rangle = 1$, $cf\langle \tau_i, \sigma_i \rangle = \tau_i \cdot \sigma_i \bmod 2$

$\qquad\qquad\qquad \xrightarrow{\quad \sigma_0, \sigma_1 \quad} \qquad\qquad \sigma_0, \sigma_1$

$w_i \in G(q)(1 \le i \le k)$, $w \equiv \displaystyle\sum_{i=1}^{k} w_i \bmod q$

$t_i \equiv g^{w_i} \bmod p (1 \le i \le k)$ $\qquad \xrightarrow{\quad t_i \quad} \qquad t_i$

$y_i^e \equiv \begin{cases} y_i^0 \equiv (s_i \cdot \sigma_0 + w_i) \bmod q \\ y_i^1 \equiv (s_i + w_i \cdot \sigma_1) \bmod q \end{cases} (1 \le i \le k)$

$\qquad\qquad\qquad \xrightarrow{\quad y_i^e \quad} \qquad y_i^e$

$$t^{(\sigma_e)^{1-e}} \cdot \prod_{i=1}^{k} t_i^{(\sigma_e)^e} \equiv \prod_{i=1}^{k} g^{y_i^e} \bmod p$$

Fig. 1. Proposed Non-interactive Authentication Algorithm

To minimize the transmitted data and optimize the performance, we modify proposed non-interactive scheme. In proposed authentication protocol, we compute $\alpha_0 \equiv g^{z_0} \bmod p$ and $\alpha_1 \equiv g^{z_1} \bmod p$ which can remove w_i in previous scheme. And the prover's secret s_i is only composed by two secret s_0 and s_1. We can calculate the *response* y_i^e $(0 \le i \le 1)$ as an equation (2).

$$y_i^e \equiv \begin{cases} y_i^0 \equiv (s_i \cdot \sigma_0 + z_i) \bmod q \\ y_i^1 \equiv (s_i + z_i \cdot \sigma_1) \bmod q \end{cases} (1 \le i \le k) \tag{2}$$

The Fig. 2 represents the optimized non-interactive scheme with modified *Schnorr* authentication based on the oblivious transfer protocol *(Sch-NIOT)*.

Prover(A)   Verifier(B)

$s_0, s_1 < p-1,\ s \equiv (s_0 + s_1) \bmod q$ $t_i \equiv g^{s_i} \bmod p (0 \le i \le 1)$

$\cdots \cdots$

$\alpha_0 \equiv g^{z_0} \bmod p$, $\alpha_1 \equiv g^{z_1} \bmod p$

$$\xrightarrow{\quad \alpha_0, \alpha_1, \sigma_0, \sigma_1 \quad} \qquad \alpha_0, \alpha_1, \sigma_0, \sigma_1$$

$y_i^e \equiv \begin{cases} y_i^0 \equiv (s_i \cdot \sigma_0 + z_i) \bmod q \\ y_i^1 \equiv (s_i + z_i \cdot \sigma_1) \bmod q \end{cases} (1 \le i \le k)$

$$\xrightarrow{\qquad y_i^e \qquad} \qquad\qquad y_i^e$$

$$\alpha_i^{x_i} \equiv \tau_i \bmod p$$

$$\prod_{i=0}^{1} t_i^{(\sigma_e)^{1-e}} \cdot \alpha_i^{(\sigma_e)^e} \equiv \prod_{i=0}^{1} g^{y_i^e} \bmod p$$

Fig. 2. *Schnorr*-type Non-interactive Authentication Algorithm (*Sch-NIOT*)

We can also improve this scheme into the *ElGamal*-type as a Fig. 3. This *ElGamal* based non-interactive scheme *(ElG-NIOT)* provides a simple verification steps on the prover's identification.

Prover(A)   Verifier(B)

$\cdots \cdots$

$y_i \equiv (s_i \cdot \alpha_i + t_i \cdot z_i) \bmod q (0 \le i \le 1)$

$$\xrightarrow{\qquad y_i \qquad} \qquad\qquad y_i$$

$$\alpha_i^{x_i} \equiv \tau_i \bmod p$$

$$g^{y_i} \equiv t_i^{\alpha_i} \cdot \alpha_i^{t_i} \bmod p$$

Fig. 3. ElGamal-type Non-interactive Authentication Algorithm (*ElG-NIOT*)

4   Applications of Proposed Non-interactive Schemes

Proposed *ElGamal*-type non-interactive scheme can be applicable to the zero knowledge proofs system using the shared strings $\sigma_0, \sigma_1 \in \{0,1\}^k (k = |p|)$. When we define σ_i^j as a j-th bit of σ_i, we can generate the zero knowledge proof $\Pr_i$ by using these shared strings. The generation algorithm is as follow Fig. 4 on the condition $C_i \in \{0,1\}$ with common proof generation function f_{PG}.

$$for \ \ i = 1,...,m$$
$$for \ \ j = 1,...,n$$
$$if \ \sigma_i^j \equiv C_0 \ then$$
$$y_i^j \equiv f_{PG}(\sigma_0^j);$$
$$if \ \sigma_i^j \equiv C_1 \ then$$
$$y_i^j \equiv f_{PG}(\sigma_1^j);$$
$$\Pr_i = (y_i^1, y_i^2,..., y_i^n)$$
$$(\Pr_1,...,\Pr_m)$$

Fig. 4. Generation of zero knowledge proofs

The prover computes his non-interactive response using the bit field of shared references. And he repeats those steps until the end of shared strings. The prover sends his proofs to the verifier. And the verifier also references the shared strings for proofing remote entity's authentication.

Proposed non-interactive zero knowledge scheme provides a simple verification steps and has an advantage in constructing the sharing of common strings than existing algorithms such as [20,21]. But if the size of shared reference is short, the security of proposed scheme is not so good as that of [21].

Agent is the automatic software for doing the necessary job behalf of user[25,26]. An agent system can provide the enhanced facilities by its integrated property of independent function based on its autonomous framework. We can use this proposed scheme in mobile entity authentication. When a mobile object, such as mobile agent in current mobile computing or mobile user in real mobile communication systems, visits one site for getting useful information or receiving wireless services, it can store those shared strings in advance. In other cases, trusted center allocates unique strings to this object. And then it can authenticate his identity by sending his own proofs without any required real-time computations. Because those shared strings have relations with his own randomly selected secrets, $\alpha_0 \equiv g^{z_0} \bmod p$ and

$\alpha_1 \equiv g^{z_1} \bmod p$, proposed NIZK scheme is non-transitive in multiple user and large distributed computing environments. The detailed algorithm of *ElG-NIZK* scheme is as follow Fig. 5.

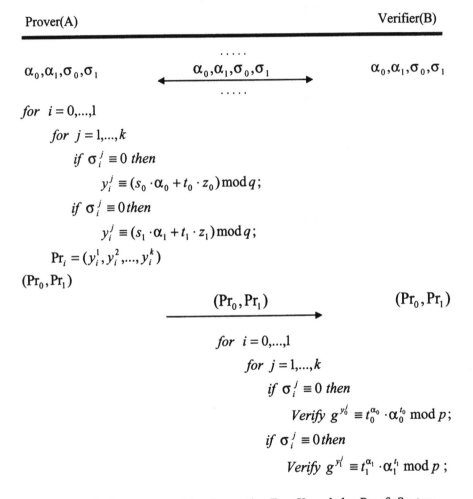

Fig. 5. Modified *ElGama*-type Non-Interactive Zero Knowledge Proofs System
(*ElG-NIZK*)

Using these shared reference strings of the oblivious transfer protocol, the prover can select the *ElGamal*-type y_i^j and construct the proof Pr_i. We can verify the integrity of these proofs as a (Proof-1).

(Proof-1) Proof of $g^{y_i^j} \equiv t_i^{\alpha_i} \cdot \alpha_i^{t_i} \bmod p$

$$if \ i = 0 \ then \ g^{y_0^j} \equiv g^{s_0 \cdot \alpha_0 + t_0 \cdot z_0} \equiv (g^{s_0})^{\alpha_0} \cdot (g^{z_0})^{t_0} \equiv t_0^{\alpha_0} \cdot \alpha_0^{t_0} \bmod p$$

$$if \ i = 1 \ then \ g^{y_i^j} \equiv g^{s_1 \cdot \alpha_1 + t_1 \cdot z_1} \equiv (g^{s_1})^{\alpha_1} \cdot (g^{z_1})^{t_1} \equiv t_1^{\alpha_1} \cdot \alpha_1^{t_1} \bmod p$$

We can proof the non-interactive properties in proposed *ElG-NIZK* scheme. We first represent both the prover and the verifier as $P = (P_1, P_2)$, $V = (V_1, V_2)$. P_1 is the information about the trusted center T. P_2 is the prover himself. Let the size k of shared reference is sufficiently large, we can check the properties of non-interactive zero knowledge proofs[20,21].

- Preprocessing step - In *ElG-NIZK* scheme, P_1 interacts with V_1 and outputs p, S_P, S_V. We assume that the shared information between P_1 and V_1 is p. The secret of P_1 and V_1 is S_P and S_V respectively in equation (3).

$$p = \sigma_i \ (0 \leq i \leq 1), \ S_P = s_i \ (0 \leq i \leq 1), \ S_V = \phi \tag{3}$$

- Steps of knowledge proofs - We can verify whether the proposed *ElG-NIZK* scheme satisfies the completeness and soundness[20,21,22,23] conditions from P_2 to V_2 without bi-directional interactions.
- Completeness: Let the random $x = \sigma_i^j$ when $x \in L \cap \{0,1\}^k$, *ElG-NIZK* scheme satisfies this equation (4) from P_2 to V_2.

$$\Pr ob(V_2(1^k, p, S_V, x, \beta) = accept : (p, S_P, S_V) \leftarrow (P_1 \leftrightarrow V_1)(1^k); \tag{4}$$
$$\beta \leftarrow P_2(1^k, p, S_P, x)) \geq 1 - 2^{-2k}$$

- Soundness: Let the random $x = \sigma_i^j$, $x \in \dot{L} \cap \{0,1\}^k$ in the interactive protocol with an operation protocol $\dot{P}_2$, *ElG-NIZK* scheme satisfies this equation (5).

$$\Pr ob(V_2(1^k, p, S_V, x, \beta) = accept : (p, S_P, S_V) \leftarrow (P_1 \leftrightarrow V_1)(1^k); \tag{5}$$
$$\beta \leftarrow \dot{P}_2(1^k, p, S_P, x)) \geq 2^{-2k}$$

In such growing areas as remote applications and wireless mobile computing system based on the large public networks, electronic commerce, digital signature and copyright protection using digital watermarking technique[30], we need to prevent the user from tampering critical data owned by those applications. Overall systems have not enough processing power and secure primitives. However, proposed schemes can also be applicable to the copyright protection systems[28,29] by using presented simple, efficient and secure non-interactive protocols in an open and insecure network environment. There are wide variety of multimedia applications that can benefit from using those suggested secure protocols such as commercial pay-per-view video multicast, audio files and on-line digital products. Proposed protocol can run in open network as it provides advanced communication bandwidth and security in efficient manner. It can provide copyright protection for the information provider without leaking his own information from the attacks.

5   Security and Performance Evaluations

Proposed schemes are also based on the intractability of the discrete logarithm problem, and it is similar with existing interactive protocols. But proposed *Sch-NIOT* scheme generalizes the interactive *Schnorr*-type authentication protocol by using k number of s_i and w_i. Suggested *Sch-NIOT* protocol will be *Schnorr*-type or *Okamoto*-type if k is 1 or 2 respectively. So, its security is same as those protocols. However, as the proposed scheme sends *non-interactive response* e with verifier's c, the computed *response* y_i^e provides a lot of complexity. On the other hand, it can upgrade the security of authentication scheme better than that of the existing interactive *Schnorr* protocol. *Sch-NIOT* scheme applies oblivious transfer protocol on the modified *Schnorr* protocol. Using *OT*, we can share the common reference strings $\sigma_i\ (0 \le i \le 1)$ between the prover and the verifier. For optimization, we use only two secret keys, s_1 and s_2, which enhances the security of *Sch-NIOT* into that of *Okamoto* protocol. Proposed scheme can verify the prover such as mobile entity by using mono-directional non-interactive proofs system. We can check the integrity of oblivious transfer protocol as a (Proof-2).

(Proof-2) Proof of $\alpha_i^{x_i} \equiv \tau_i \bmod p$ on the oblivious transfer protocols.

$$\text{if } i = 0 \text{ then } \beta_0 = g^{x_0}, \beta_1 = c \cdot (g^{x_0})^{-1}$$

$$\alpha_i^{x_i} \equiv \alpha_0^{x_0} \equiv (g^{z_0})^{x_0} \equiv (g^{x_0})^{z_0} \equiv (\beta_0)^{z_0} \equiv \tau_0 \bmod p$$

$$\text{if } i = 1 \text{ then } \beta_1 = g^{x_1}, \beta_0 = c \cdot (g^{x_1})^{-1}$$

$$\alpha_i^{x_i} \equiv \alpha_1^{x_1} \equiv (g^{z_1})^{x_1} \equiv (g^{x_1})^{z_1} \equiv (\beta_1)^{z_1} \equiv \tau_1 \bmod p$$

And we can verify the integrity of non-interactive authentication scheme as follow (Proof-3).

(Proof-3) Proof of $\displaystyle\prod_{i=0}^{1} t_i^{(\sigma_e)^{1-e}} \cdot \alpha_i^{(\sigma_e)^e} \equiv \prod_{i=0}^{1} g^{y_i^e} \bmod p$

$$\text{if } e = 0 \text{ then}$$

$$\prod_{i=0}^{1} t_i^{(\sigma_e)^{1-e}} \cdot \alpha_i^{(\sigma_e)^e} \equiv \prod_{i=0}^{1} t_i^{(\sigma_0)^1} \cdot \alpha_i^{(\sigma_0)^0} \equiv \prod_{i=0}^{1} g^{s_i \cdot \sigma_e} \cdot \alpha_i$$

$$\equiv g^{(s_0+s_1)\sigma_0} \cdot \alpha_0 \cdot \alpha_1 \equiv g^{s \cdot \sigma_0} \cdot \alpha_0 \cdot \alpha_1 \equiv t^{\sigma_0} \cdot \alpha_0 \cdot \alpha_1$$

$$\prod_{i=0}^{1} g^{y_i^e} \equiv \prod_{i=0}^{1} g^{y_i^0} \equiv \prod_{i=0}^{1} g^{s_i \cdot \sigma_0 + z_i} \equiv g^{s_0 \cdot \sigma_0 + z_0} \cdot g^{s_1 \cdot \sigma_0 + z_1}$$

$$\equiv g^{(s_0+s_1)\sigma_0} \cdot g^{(z_0+z_1)} \equiv g^{s \cdot \sigma_0} \cdot g^{(z_0+z_1)} \equiv t^{\sigma_0} \cdot \alpha_0 \cdot \alpha_1$$

if e = 1 then

$$\prod_{i=0}^{1} t_i^{(\sigma_e)^{1-e}} \cdot \alpha_i^{(\sigma_e)^e} \equiv \prod_{i=0}^{1} t_i^{(\sigma_1)^0} \cdot \alpha_i^{(\sigma_1)^1} \equiv \prod_{i=0}^{1} t_i \cdot \alpha_i^{\sigma_1}$$

$$\equiv g^{(s_0+s_1)} \cdot (\alpha_0 \cdot \alpha_1)^{\sigma_1} \equiv g^s \cdot (\alpha_0 \cdot \alpha_1)^{\sigma_1} \equiv t \cdot (\alpha_0 \cdot \alpha_1)^{\sigma_1}$$

$$\prod_{i=0}^{1} g^{y_i^e} \equiv \prod_{i=0}^{1} g^{y_i^1} \equiv \prod_{i=0}^{1} g^{s_i + z_i \cdot \sigma_1}$$

$$\equiv g^{s_0 + z_0 \cdot \sigma_1} \cdot g^{s_1 + z_1 \cdot \sigma_1} \equiv g^s \cdot g^{z_0 \cdot \sigma_1} \cdot g^{z_1 \cdot \sigma_1} \equiv t \cdot (\alpha_0 \cdot \alpha_1)^{\sigma_1}$$

Proposed *ElG-NIOT* scheme uses similar public/private key pair. Only in a generation of *response*, *ElG-NIOT* applies the *ElGamal*-type protocol. When the system parameter k is 1, suggested scheme is similar with existing *ElGamal* protocol. So, the security of *ElG-NIOT* is same as that of *ElGamal*[14] protocol. And its safety is more enhanced in case of $k \geq 2$.

ElG-NIOT scheme converts the *ElGamal* based interactive protocol into the non-interactive scheme. As proposed scheme is based on the *ElGamal* protocol, the total amount of computation can be diminished in the verification step on the prover's y_i. Therefore, proposed *ElG-NIOT* scheme is compatible to the mobile object based computing which is bounded on the computation and its communication bandwidth. For verifying the integrity of *ElG-NIOT* scheme, we first prove the τ_0, τ_1 and α_0, α_1 received from the oblivious transfer protocol. And then we can verify the *ElGamal*-type non-interactive authentication protocol as (Proof-4).

(Proof-4) Proof of $g^{y_i} \equiv t_i^{\alpha_i} \cdot \alpha_i^{t_i} \bmod p$

$$g^{y_i} \equiv g^{s_i \cdot \alpha_i + t_i \cdot z_i} \equiv (g^{s_i})^{\alpha_i} \cdot (g^{z_i})^{t_i} \equiv t_i^{\alpha_i} \cdot \alpha_i^{t_i} \bmod p$$

In proposed protocols, as the prover generates his own random number, a forger can't get any information without knowing it. Thus the calculation of appropriate challenge seems to be as difficult as the computation of the discrete logarithm. Additionally, as the values used in the initial parameters of oblivious transfer protocol are secretly selected, we can securely verify its correctness in entity authentication process. We can assume that the security of the proposed non-interactive scheme is similar with that of *ElGamal* signature scheme.

The security analysis for a total break of the authentication scheme and universal forgery of messages can also be adapted from these schemes. Proposed scheme uses oblivious transfer protocol and generates private parameters, $\beta_i \equiv g^{x_i} \bmod p$ and $\beta_{1-i} \equiv c \cdot (g^{x_i})^{-1} \bmod p$. An attacker can choose signature parameters at random and calculate the corresponding values. To avoid a total break of the proposed scheme, the forger can randomly choose the secrets $z_0, z_1 \in \{0, ..., p-2\}$, then attempts to find $\sigma_0, \sigma_1 \in \{0,1\}^k (k = |p|)$. In this method, the forger must solve the discrete logarithm of $\tau_0 \equiv \beta_0^{z_0} \bmod p$ and $\tau_1 \equiv \beta_1^{z_1} \bmod p$. These computations are extremely difficult and it's still an open question whether it is more difficult than solving the discrete logarithm problem. Additionally, proposed protocols use divided and shared secrets $s_i < p-1 (1 \le i \le k)$ in private key generation steps. This mechanism is useful for enhancing overall security of proposed scheme although it brings disadvantages in aspect of total number of modular computation.

Proposed schemes are compared with existing non-interactive oblivious transfer protocol and zero knowledge proofs system. Table 1 shows the results of this performance evaluation.

Table 1. Performance Evaluation in Non-Interactive Schemes

Items / Methods	NIOT[19]	Sch-NIOT	EIG-NIOT	NIZK[21]	EIG-NIZK
Primitive Problem	Discrete Logarithm OT	Discrete Logarithm OT-Schnorr	Discrete Logarithm OT-ElGamal	Discrete Logarithm ZK	Discrete Logarithm ZK-ElGamal
ID-based Variant	•	•	•		•
Primary Function	Oblivious Transfer	Schnorr based Auth.	ElGamal based Auth.	Zero Knowledge Proofs	ElGamal based ZK Proofs
Mobile Entity Authentication	×	•	•	•	•
Fair Cryptosystem ?	×	•	•	•	•
Non-Interactive Proofs	•	•	•	•	•
Size of Shared Strings(bits)	2k	2k	2k	k(n+1)	2k
Communication(bits)	1024+2k	1816+2k	1816+2k	mk(n+1)	280k
Preprocessing(Prover)	2p	2p	2p	mkp	2kp
On-line Processing(Prover)	0	0	0	0	0
On-line Processing(Verifier)	2p	2p	2p	mkp	2kp
Total Processing 512-bits Modular Multi.	4p	4p	4p	2mkp	4kp
• : Yes(possible), • : Not Known, × : No(impossible), cf. m, k, p, n : System Parameters					

As we pointed out earlier, proposed non-interactive schemes have secure primitives as they are similar with the traditional protocol based on the discrete logarithm problems. Therefore, these schemes can also be applicable to an entity authentication protocol needed in mobile computing environments and digital content providing system.

The size of reference strings in normal non-interactive proofs is $k(n+1)$ and another is $2k$. In bandwidth, proposed *Sch-NIOT* and *ElG-NIOT* schemes have to send more bits. But we can convert these schemes into more advanced ones. Comparing with the *NIZK*, proposed *ElG-NIZK* scheme enhances the communication bandwidth. The *OT* based schemes have a better performance in the aspects of processing.

6   Conclusions

There are diverse efforts for enhancing the security of multimedia content, on-line digital product system combined with electronic payment system based on the several cryptographic protocols. Secure frameworks in copyright protection and distributed digital contents providing system require precise authentication among its entities. But existing protocols are based on the bi-directional interactive transactions based on the traditional client/server models. Therefore, interactive based authentication schemes have a disadvantage in its bandwidth on overall systems.

In this study, we propose non-interactive entity authentication schemes suitable for diverse applications in the aspects of security and performance. We evaluate the existing interactive protocols. Then, they are converted into the non-interactive schemes using oblivious transfer protocol. Suggested schemes provide advanced safety and performance on authenticating each entity as these transactions are done with minimized interaction in mono-directional manner. Proposed non-interactive authentication schemes are applicable to diverse applications such as distributed electronic commerce and access control, etc. We can develop those schemes into digital signature and combine steganographic primitives for non-interactive copyright protection. Especially, proposed schemes can be advanced into the publicly verifiable protocols on the multimedia products by using the secret sharing primitives on private copyright information.

References

1. Stefan Brands, "Off-Line Cash Transfer by Smart Card," CWI Report CS-R9455, (1994)
2. Stefan Brands, "Electronic Cash on the Internet," Symposium of Network and Distributed System Security, (1995)
3. M. Burrows, M. Abadi, R. Needham, "The Scope of a Logic of Authentication," Technical report, DEC System Research Center, (1994)
4. Gustavus J. Simmons, "A Survey of Information Authentication," Proceedings of the IEEE, Vol. 76, No. 5, (1988) 603-620
5. W. Diffie and M. Hellman, "New Directions in Cryptography," IEEE Transactions on Information Theory, Vol. IT-22, No. 6, (1976) 472-492
6. Whitfield Diffie, "The First Ten Years of Public-Key Cryptography," Proceedings of the IEEE, Vol. 76, No. 5, (1988) 560-577

7. C. P. Schnorr, "Efficient Identification and Signatures for Smart Cards," Advances in Cryptology, Proceedings of Crypto'89, Springer-Verlag, (1990) 239-252
8. Tatsuaki Okamoto, "Provably Secure and Practical Identification Schemes and Corresponding Signature Schemes," Advances in Cryptology, Proceedings of Crypto'92, Springer-Verlag, (1993) 31-53
9. B. Schneier, Applied Cryptography, 2nd Edition, John Wiley & Sons Press (1996)
10. Alfred J. Menezed, Paul C. van Oorschot, Scott A. Vanstone, Handbook of Applied Cryptography, CRC Press (1996)
11. A. Fiat and A. Shamir, "How to Prove Yourself: Practical Solution to Identification and Signature Problems," Advances in Cryptology, Proceedings of CRYPTO'86, Springer-Verlag, (1987) 186-199
12. U. Feige, A. Fiat, A. Shamir, "Zero Knowledge Proofs of Identity," Proceedings of the 19th Annual ACM Symposium of Theory of Computing, (1989) 210-217
13. L. C. Guillou and J. J. Quisquater, "A Practical Zero-Knowledge Protocol Fitted to Security Microprocessor Minimizing Both Transmission and Memory," Advances in Cryptology, Proceedings of Eurocrypt'88, Springer- Verlag, (1989) 123-128
14. T. ElGamal, "A Public Key Cryptosystem and a Signature Scheme based on Discrete Logarithms," IEEE Transactions on Information Theory, Vol. IT-30, No. 4, (1985) 469-472
15. R. L. Rivest, A. Shamir and L. Adleman, "A Method Obtaining Digital Signatures and Public-Key Cryptosystems," Communications of the ACM, Vol. 21, No. 2, (1978) 120-126
16. S. Goldwasser, S. Micali, C. Rackoff, "The Knowledge Complexity of Interactive Proofs," SIAM Journal of Computing, Vol. 18, No. 1, (1989) 186-208
17. Kazuo Ohta and Tatsuaki Okamoto, "A Modification of the Fiat-Shamir Scheme," Advances in Cryptology, Proceedings of Crypto'88, Springer-Verlag, (1989) 232-243
18. M. Rabin, "How to exchange secrets by oblivious transfer," Technical Reports TR-81, Harvard Aiken Computation Laboratory, (1981)
19. Mihir Bellare, Silvio Micali, "Non-Interactive Oblivious Transfer and Applications," Advances in Cryptology, Proceedings of Crypto 89, Springer-Verlag, (1989)
20. M. Blum, P. Feldman, s. Micali, "Non-Interactive Zero-Knowledge Proof Systems and Applications," Proceedings of the 20th Annual ACM Symposium on Theory of Computing, (1988)
21. Alfredo De Santis, Giovanni Di Crescenzo, Pino Persino, "Randomness-Efficient Non-Interactive Zero Knowledge," ICALP'97 Conference, (1997)
22. A. D. Santis, S. Micali, G. Persiano, "Non-Interactive Zero-Knowledge Proof Systems," Advances in Cryptology, Proceedings of Crypto'87, Vol. 293, (1988)
23. A. D. Santis, S. Micali, G. Persiano, "Non-Interactive Zero-Knowledge Proof-Systems with Preprocessing," Advances in Cryptology, Proceedings of Crypto'88, Vol. 403, (1989)
24. S. Micali, "Fair Cryptosystems," Technical Reports MIT/LCS/TR-579-b, (1993)
25. R. Gray, D. Kotz, S. Nog, D. Rus and G. Cybento, "Mobile Agents for Mobile Computing," Proc. Of 2nd Aizu Int'l Symposium on Parallel Algorithm/Architecture Synthesis(pAs97), Fukushima, Japan, Mar., (1997)
26. C. G. Harrison, D. M. Chess and A. Kershenbaum, "Mobile Agents : Are they a good idea?" Technical Report, IBM T.J. Watson Research Center, Mar., (1995)
27. T. Gilmont, J.-D. Legat, J, -J. Quisquater, "Architecture of Security Management Unit for Safe Hosting of Multiple Agents," Security and Watermarking of Multimedia Contents, Proceedings of SPIE, Vol. 3657, (1999)
28. E. van Faber, R. Hammelrath, FP Heider, "The Secure Distribution of Digital Contents," ACSAC'97, (1997) 16-22
29. N. Memon, PW Wong, "Protecting Digital Media Content," Communications of the ACM, Vol. 41, No. 7, (1998) 35-43
30. G. Voyatzis, N. Nikolaidis, I. Pitas, "Digital Watermarking: an Overview," EUSIPCO'98, (1998) 9-12

Implementation of Virtual Private Networks at the Transport Layer

Jorge Davila[1], Javier Lopez[2], and Rene Peralta[3]

[1] Computer Science Department, Facultad de Informatica
Universidad Politecnica de Madrid, 28660 - Madrid
jdavila@fi.upm.es
[2] Computer Science Department, E.T.S. Ingenieria Informatica
Universidad de Malaga, 29071 - Malaga
jlm@lcc.uma.es
[3] Center for Cryptography, Computer, and Network Security (CCCNS)
University of Wisconsin-Milwaukee, P.O. Box 784, Milwaukee - WI 53201
peralta@cs.uwm.edu

Abstract. Virtual Private Network (VPN) solutions mainly focus on security aspects. Their main aims are to isolate a distributed network from outsiders and to protect the confidentiality and integrity of sensitive information traversing a non-trusted network such as the Internet. But when security is considered the unique problem, some collateral ones arise. VPN users suffer from restrictions in their access to the network. They are not free to use traditional Internet services such as electronic mail exchange with non-VPN users, and to access Web and FTP servers external to the organization. In this paper we present a new solution, located at the TCP/IP transport layer that, while maintaining strong security features, allows the open use of traditional network services. The solution does not require the addition of new hardware because it is an exclusively software solution. As a consequence, the application is totally portable. Moreover, the implementation is located at the transport layer; thus, there is no need to modify any software previously installed, like FTP, Telnet, HTTP, electronic mail or other network applications.

1 Introduction

Evolution of commercial needs and transference of company resources to computers have forced organizations to interconnect the networks of their branch offices, setting up their own private networks.

Traditionally, companies have used leased lines with that purpose. The most representative example is *Frame-Relay* service [2,9], which is based on the transfer of information frames between intermediate switching offices. The service, that uses *permanent virtual circuits* (PVCs) through telephone network routers, presents some drawbacks:

M. Mambo, Y. Zheng (Eds.): ISW'99, LNCS 1729, pp. 85–102, 1999.

- It becomes expensive because connections remain open permanently.
- The architecture creates large latency periods because of the poor connectivity between intermediate routers.
- Full connectivity requires the increment of PVCs and, hence, of intermediate network routers; but the cost of avoiding routing problems in this way is high.
- The number of companies that offer Frame-Relay services is small compared to the number of *Internet Service Providers* (ISPs), so competitiveness is more limited.

On the other hand, *open networks* offer a more profitable solution than leased lines. Thus, for example, *Virtual Private Networks* (VPNs) use relatively low-cost, widely available access to public networks to connect remote sites together safely. Network architectures defined by VPNs are inherently more scalable and flexible than classical WANs, and they allow organizations to add and remove branch offices into their systems in an easy way.

However, and as shown later, the study of the different TCP/IP [19] stack layers reveals that the different solutions that enable establishing a VPN essentially focus on security aspects. Their main aims are to isolate a distributed network from outsiders and to protect the privacy and integrity of sensitive information traversing the non-trusted open networks, as the Internet. These approaches fail to be complete.

The main drawback in conceiving the security problem as the unique target is that VPN users suffer from restrictions in accessing the Internet. That is, they cannot freely use traditional services such as electronic mail exchange with non-VPN users, and cannot freely access Web and FTP servers external to the organization. Actually, within the same application, it is a difficult task to enable generic Internet access to VPN users and, at the same time, to provide a strong enough security model for the organization.

This work presents an approach to this problem. We have developed a new security solution for VPNs that, while maintaining strong security features, allows the open use of traditional Internet services. The solution does not require the addition of new hardware because it is an exclusively software solution. As a consequence, the application is totally portable. Moreover, the implementation is located at the transport layer; thus, there is no need to modify any software previously installed, like FTP, Telnet, electronic mail or other network applications.

The paper is organized in the following way: Section 2 introduces the different cryptographic solutions used in the TCP/IP stack layers to establish private network communications over the Internet, focusing on their advantages and disadvantages. Section 3 explains the two-module architecture of the new system. Section 4 describes the operation of the first module, called *Secsockets*, an extension of the traditional socket interface to which security features have been added. Section 5 outlines the operation of *VPN-Insel*, the second module. This is the part of the system that really establishes the VPN, and resides on the interface of the previous module. Section 6 shows a real example of how the new solution works, and section 7 presents concluding remarks.

2   Overview of VPN Solutions in the TCP/IP Stack Layers

There are different cryptographic solutions in the TCP/IP stack layers that can be used to establish confidential and authenticated communications over Internet (figure 1). In this section we briefly analyze the most relevant solutions in those layers, pointing out their advantages and disadvantages.

Link Layer

The method used in the lowest stack layer is point-to-point encryption, which is the classic method of using cryptography for digital communications in networks of static geometry.

The element used to encrypt at link level is the *in-line encryptor*. This is a hardware device with two data ports: one always handles plaintext and the other always handles ciphertext [7]. When plaintext data arrive on the plaintext port, the encryptor transforms data into ciphertext and transmits them out to the ciphertext port. Data are likewise transformed from ciphertext to plaintext after arriving on the ciphertext port.

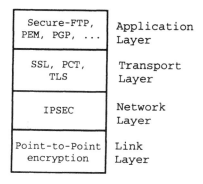

Fig. 1. Cryptographic solutions in the TCP/IP stack layers

The architecture of link level encryption defines a completely isolated connection between two systems. On the one hand, physical access is restricted because every connected host belongs to the organization; on the other hand, logical access is restricted too because information is encrypted throughout the transmission. This solution does not necessarily require the use of the TCP/IP stack protocols because these ones work at higher levels.

Point-to-point encryption is a simple concept that makes it, from a security standpoint, a good VPN solution. However, it has certain properties that make it hard to use in every application. Firstly, system users have no choice regarding encryption. They can either link to a host using encryption or they cannot communicate at all. Therefore, it does not provide generic Internet access because it protects from hostile outsiders, but also blocks access to beneficial outsiders. Secondly, point to point encryption scales very poorly. It becomes complex and expensive to establish links to new hosts that are installed in the network system.

Network Layer

The solution used at the network layer is the encryption of the data fields in the IP packets. The set of IPSEC protocols (IP Security Protocols) [1], is a part of IPng (or IPv6), the IP's future version and it is designed to provide privacy and/or data forgery detection.

For this purpose, IPSEC defines two optional packet headers: *Authentication Header* (AH), and *Encapsulating Security Payload* (ESP). Both headers contain a numeric value called the *Security Parameter Index* (SPI), which is used by a host to identify the cryptographic keys and the security procedures to be used.

In order to communicate, each pair of hosts using IPSEC must negotiate a *security association*. The security association establishes what types of protection to apply, how to encrypt or authenticate, and which keys are needed. The packet IP address and the packet SPI in the packet header determine the security association applied to an IPSEC header.

The combination of IPSEC protocols with routers or firewalls constitutes a VPN solution, and it enables the use of traditional Internet services. Nevertheless, and because of its location, it is difficult to deliver a security relationship between two applications, which makes difficult to provide the same granularity and control over authentication and encryption. Subsequently, an important disadvantage of this solution is that the system automatically applies protection according to the security associations between the host; users cannot decide when to apply security mechanisms, which is inefficient for most cases. An additional disadvantage is that this architecture blocks communications to other non-trusted computer systems and networks, so it does not enable generic Internet access.

Transport Layer

There are some specific solutions at the transport layer. The implementation of Netscape's *Secure Socket Layer* (SSL) [14] protocol stands out from other solutions. It is widely used in conjunction with World Wide Web service and includes capabilities for authentication, encryption and key exchange. Some similar alternatives to this solution are *Transport Layer Security* (TLS) [4], proposed by the IETF TLS working group, and the Microsoft compatible protocol *Private Communication Technology* (PCT) [13].

The SSL protocol is integrated in both Web client and server software, and protects data transfer between them. Client and server negotiate the type of protection by setting up a group of cryptographic parameters that includes a strong encryption algorithm and a shared secret key. A public key cryptosystem is used to facilitate authentication and distribution of the secret key.

Applying security at the transport level in this way provides good control over the mechanisms to be used, because a Web browser can choose whether a particular connection will be secure or not.

However, a serious disadvantage of this scheme is that it requires the integration of SSL protocol software into the application itself; that is, it is necessary to modify every network application that needs to use these security mechanisms. Therefore, this is not a general solution for a VPN because it only solves specific network services.

Application Layer

There are also some particular solutions at the application layer. This is so for electronic mail combined with *Pretty Good Privacy* (PGP) [20] or *Privacy Enhanced Mail* (PEM) [12]. Both of these apply a diversity of cryptographic solutions to electronic mail by using PGP or PEM software packages, respectively. Apart from these solutions, more software is being developed to add security to other specific Internet services: Secure-FTP [10], S-HTTP [17], S/MIME [15],etc.

The main drawback of this type of solution is that it would be necessary to guarantee that all users update every network application installed in their computer systems, replacing them with those that also provide security features. This is hardly feasible. Even if this was the case, this solution would not be homogeneous because each application would use different encryption algorithms, digital signature algorithms, hash functions, key lengths, etc., possibly causing integrity problems and holes in the system. VPNs security must be discussed in terms of networks interconnection points, and not as attributes of the elements that constitute them.

Therefore, homogeneous control of all network services is necessary to gain higher user confidence regarding the entire system. Such homogeneous control is provided by the solution we have developed.

3   Outline of the New Scheme

The scheme we introduce is an entirely software solution that works at the transport level. The solution adopts the advantages of IPSEC because it facilitates the use of traditional Internet services. However, the new design presents notable differences with IPSEC too because it allows users to decide when to apply security mechanisms.

It also enables access to any remote node; hence, any VPN user can connect to other VPN and non-VPN users. Additionally, it does not require the modification of the software of traditionally insecure applications such as FTP, HTTP, Telnet, electronic mail, etc. As a result, the scheme is a simple and cheap solution for those organizations that want to install a VPN.

It is necessary to point out what kind of risks are intrinsic to this approach. As stated before, the main characteristic of a VPN is its operation over an unsecured public network. From the VPN standpoint, the free connection of internal equipment to a public network implies a potential hole in the security perimeter defined by the VPN. If any VPN user can operate inside and outside the VPN, nothing would stop him/her to setup gateways without the knowledge of the organization security administrators.

The main reason to keep this ambiguity in our design can be understood considering two different scenarios: the mobile systems that connect to the VPN from different places (operating through ISPs, GSM equipment, etc.), and the characteristic LAN organization.

In the first case, it is clear that mobile systems must be part of the VPN, independently of its additional use inside the public network. It is hard to imagine marketing agents connecting to the corporate VPN in order to show data and information to their clients and, at the same time, not being able to connect to the competitors public websites to see their advertising.

In the case of a LAN, it is difficult that any organization accepts the loss of external accessibility to some nodes when distant branch offices connect into a single VPN. But it is clear that, because of their role and information, some nodes in the VPN are purely internal to it and must not to be accessed from outside, although some others can get access to services in the public network.

Our solution tries to satisfy both requirements; that is, it is valid for mobile systems and for proxy servers connecting internal VPN nodes to public communication services. In the last case, accessibility will depend on the configuration of the proxy servers. It is easy to set that a computer system allowed to access the public network is not allowed to access other systems that contain the organization sensitive information.

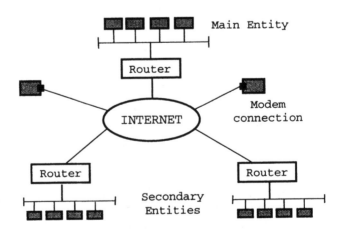

Fig. 2. Scenario for a VPN

The generic organization scenario that has been used for the development of the application is depicted in figure 2. There is a *main entity* that centralizes control of the organization and represents the Headquarters. There are other entities that represent the branch offices of the organization. These are the *secondary entities*, each of which possesses its own LAN. Finally, there are remote users who connect to the system through a dial-up connection. As it can be seen, this scenario covers the needs of any kind of organization.

Our design is divided into two sublevels, called Secsockets and VPN-Insel, respectively. Figure 3 shows their location inside the TCP/IP stack.

The module Secsockets is the lower sublevel. It is an interface that works on top of the traditional socket interface that provides access to TCP. The service offered by Secsockets is the transmission of data from an origin process to a target one through a secure and authenticated channel.

The module VPN-Insel is the upper sublevel. It uses the services provided by Secsockets and, at the same time, it supports the applications with the VPN services that these ones require. Next sections show the operation of both modules in detail.

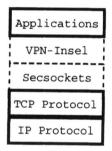

Fig. 3. Location of the new design

4   Secsockets Design and Operation

Secsockets, an extension of the traditional socket interface, enables authenticated and confidential communications. That is, Secsockets provides the same functions as the socket interface and, in addition, provides the four basic security services: authentication, confidentiality, integrity, and non-repudiation. This interface is based on a client/server architecture, in such a way that the process that starts the communication plays the role of the client, and the process that attends to the connection at the other end plays the role of the server. The operation of Secsockets is divided into two phases: a prior connection phase and the communication phase itself.

Connection Phase

Firstly, during the connection phase, mutual authentication is carried out. Secondly, there is a negotiation of the parameters (hash function, encryption algorithm, and optional compression algorithm) that will be used during the communication phase.

In this phase the two interacting parts have no secret key to share as yet. So, the use of public key cryptography [5] provides mutual authentication and confidentiality during parameter negotiation. Public key techniques enables communicating the secret key, so avoiding the risk of a third party interception, and facilitating the identification of VPN users because it provides a method to implement digital signatures.

However, public key cryptography introduces an additional problem because users have to be confident regarding other users' public keys. Public key *certificates* and the entities that issue them, *Certification Authorities* (CAs), have to be involved in this process.

There are different ways to organize the CAs of a system, giving rise to different types of *Public Key Infrastructures* (PKIs). Our solution is flexible because it permits interaction with:

- *Flat PKIs*: They only use one CA to issue user public key certificates, which is a typical option for small organizations where the number of users is not high; and
- *Hierarchical PKIs*: They use multiple CAs located in several levels. This is a suitable option for big organizations, in which the levels where the CAs are located directly correspond to the levels in the organization's structure.

The tasks performed in the connection phase are summarized in four steps:

1. Client and server exchange their certificates and validate them to obtain the respective public keys. These certificates follow the X.509 format [3].
2. The client sends a connection request to the server (figure 4). This message contains the security parameters. In order to preserve confidentiality, this message is encrypted with the server public key.
   The hash functions that can be used are MD5 [16] and SHA [8]. The symmetric key encryption algorithms that can be used are DES [6], IDEA [11] and Blowfish [18]. In case data compression is selected, the *gzip* algorithm is used.
3. Once the request is processed, the server sends a message to the client indicating acceptance or rejection. The message, which structure is depicted in figure 5, is encrypted with the client's public key.
   In case the connection is rejected, the possible error codes in *field1* are:
   - 01: version is not supported
   - 02: certificate is not valid
   - 03: hash function is not supported
   - 04: encryption algorithm is not supported

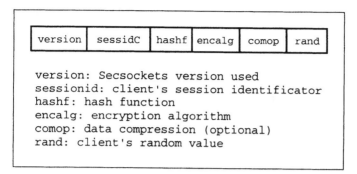

Fig. 4. Connection-request message structure

If the error code value is *01*, that is, the version is not supported, *field2* includes the version supported by the server. If the error code value is *03* or *04*, *field2* includes extra information indicating an optional choice. The information contained in that field allows the client to repeat step 2, modifying some of the parameters in order to establish the connection.

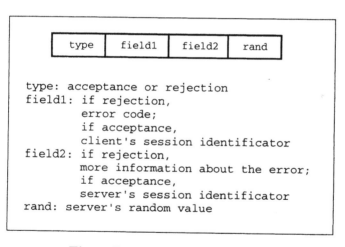

Fig. 5. Response message structure

4. Both client and server calculate the secret key that will be used during the next phase: the communication phase. Random values exchanged in step 2 and 3 and the hash function negotiated in these steps are used to calculate a bit string. The secret key is formed, according to the encryption algorithm, by selection of the appropriate number of bits. The string is also used to form a secondary key that will be used to calculate a message authentication code (MAC) for each message frame. The string is obtained from the following expression:

$$hash(hash(client\text{-}random\text{-}value) \oplus hash(sever\text{-}random\text{-}value))$$

Communication Phase

The operations performed during the communication phase (or transmission phase) are, for each message: a) fragmentation of the message into frames; b) data field compression, if this was decided during the initial negotiation; c) calculation of hash value for each frame; and, d) frame encryption and MAC calculation.

Obviously, after the reception of each frame, the receiver has to undo all previous operations. Figure 6 shows the composition of each message frame. The maximum size of the fragments depends on the packet length accepted by the network, so this is a configurable parameter.

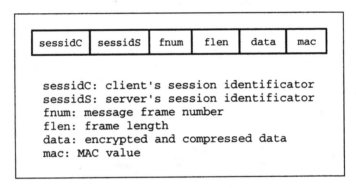

Fig. 6. Message frames structure

4.1   Functions of the Interface Secsockets

The service offered by Secsockets is to send data from a source process to a target one through a secure and authenticated channel. Secsockets works as a connection oriented communication protocol based on the client/server paradigm. For this reason, the interface provides the following group of functions:

- *sec-init()*: This function creates a connection point at the server's end. The point is initialized and the function assigns it to a local communication port. Then the server remains in an idle state waiting for a client connection-request through that port.

```
int sec_init (port, serv_addr)
      int port;                           /* Local port number */
      struct sockaddr_in *serv_addr;
```

where *struct sockaddr_in* is defined as:

```
struct sockaddr_in {
      short sin_family;       /* AF_INET: Internet protocol family */
      u_short sin_port;       /* port number (16 bits in network
```

```
                                 order byte) */
      u_long s_addr;             /* network and host identifiers
                                    (32 bits in network order byte) */
      char sin_zero[8];          /* not used */
}
```

- *sec_accept ( )*: This function is invoked by the server to positively acknowl-
  edge an incoming connection and immediately provides it with a new socket.
  After accepting the connection the negotiation phase starts at server's end.

```
int sec_accept (sockfd, cli_addr, clilen, key, param, cert,
                sessionid, r)
    int sockfd;                  /* socket identifier */
    struct sockaddr_in *cli_addr;
    int *clilen;                 /* length of cli_addr struct. */
    char *key;                   /* secret key */
    struct msgneg *param;        /* negotiated parameters */
    struct certificate cert;     /* server certificate */
    int sessionid;               /* session identifier */
    char *r;                     /* server random value*/
```

where *struct certificate* corresponds to the structure of the public-key certificate, and *struct msgneg* corresponds to the structure of the connection
request message shown in figure 4.

- *sec_connect ( )*: This function creates a final connection point at the client's
  end and starts a connection with the socket returned by the function. Afterwards, the negotiation phase starts at the client's end, authenticating the
  other end and agreeing the security parameters.

```
int sec_connect (host, port, serv_addr, key, param, cert,
                 sessionid)
    char *host;                  /* remote host address */
    int port;                    /* remote port */
    struct sockaddr_in *serv_addr;
    char *key;                   /* secret key */
    struct msgneg *param;        /* negotiated parameters */
    struct certificate cert;     /* client certificate */
    int *sessionid;              /* session identifier */
    char *r;                     /* client random value*/
```

From this point on, and following the parameters settled on during the negotiation phase, it will be possible to perform securely the communication phase.
The functions of this phase are described next:

- *sec_recv ( )*: This function enables data reception through a secure socket
  connection, decrypts them, checks their authenticity, and decompresses them
  if necessary.

```
int sec_recv (ssockfd, buffer, key, param, sessionid)
    int ssockfd;          /* socket identifier */
    char *buffer;         /* buffer for received data */
    char *key;            /* secret key */
    struct msgneg param;  /* parameters to be used */
    int sessionid;        /* session identifier */
```

- *sec_send ( )*: This function is symmetrical to the previous one. It compresses data if this was negotiated, calculates the hash value, encrypts data, and finally, sends them through the socket. These operations are performed according to the negotiations of the client and server.

```
int sec_send (ssockfd, buffer, key, param, sessionid)
    int ssockfd;          /* socket identifier */
    char *buffer;         /* buffer for data to be sent */
    char *key;            /* secret key */
    struct msgneg param;  /* parameters to be used */
    int sessionid;        /* session identifier */
```

- *sec_close ( )*: This function erases the socket.

```
int sec_close(ssockfd)
    int ssockfd;          /* socket identifier */
```

Figure 7 shows a time line of the typical scenario that takes place for the transfer.

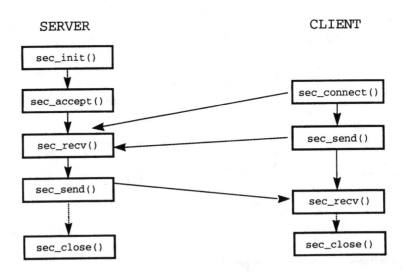

Fig. 7. Seccockets sequence calls

5   VPN-Insel Design and Operation

The module VPN-Insel uses the services provided by the interface Secsockets. At the same time it supports the communication applications that run at the upper level and provides them with the characteristic services of a VPN. So, by making use of Secsockets security mechanisms, VPN-Insel offers users the possibility of registering in the VPN, participating in on-line connections, like HTTP, Telnet or FTP, and off-line connections, such as electronic mail.

Moreover, this module manages the VPN. That is, VPN-Insel initializes the private network, setting the configuration parameters and the existing CAs. Once the private network is working, VPN-Insel controls which users belong to it and can use its services, and which new users and secondary entities can register. It also allows the system administrator to install HTTP, Telnet, FTP and SMTP servers which are for the exclusive use of users belonging to the VPN.

Basically, the VPN-Insel processes and their interrelations are those shown in figure 8. As can be seen in the figure, there are different types of server and client processes.

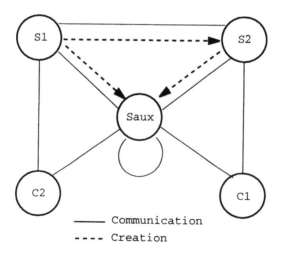

Fig. 8. Communication among VPN-Insel processes

We can distinguish three different types of server processes:

- The process symbolized by *S1* represents the first type of server, called the *main server*, because it runs in the main entity's computer system (main LAN in the organization). This type of process centralizes the VPN management, controls other types of server processes, and controls remote users. Whenever a user wants to establish a connection with the VPN, the main server checks if the user is allowed to do this by querying a database that contains users' information. There is only one of these processes in each VPN.

- One *secondary server*, represented by *S2*, exists in each entity, i.e., in each LAN of the organization. This server controls the local users (the users inside its LAN) and decides which of them can log in to the VPN. It also attends to their requests in the same way as the main server does.
- *Saux* represents the third type of process, the *auxiliary server*. Auxiliary servers are created by the main server or by a secondary one to attend to users during a particular connection. They are intermediate processes within secure communications and make use of the functions that the Secsockets interface provides, in such a way that security mechanisms are transparent to client processes.

As regards client processes these are divided into two categories *C1* and *C2*:

- Type *C1* represents the client software for those users who communicate from a LAN.
- Type *C2* is a client process specialized for remote users. These ones are not located in a LAN of the VPN, so their software does not include secondary servers processes. For this reason the client process has to directly manage the security function. This type of client process has a friendly interface that facilitates the work of final users.

As previously mentioned, our design considers two kinds of communications. The first one is LAN to LAN, and the second one is remote-user to LAN. Both alternatives are depicted in figures 9 and 10, respectively.

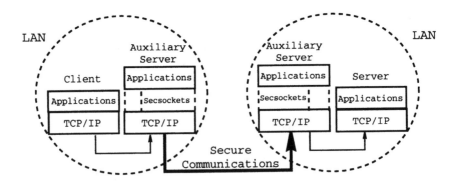

Fig. 9. LAN to LAN communication under VPN-Insel

In the case of LAN to LAN communication, there are two auxiliary servers, one on each end of the line. These work as routers, receiving packets through the

Secsockets interface and communicating using the client software of the VPN-Insel module. In this way, communication between both auxiliary servers, i.e., the stretch across the Internet, is secure.

As it can be seen, the role of an auxiliary server is that of an intermediary program between an external server and an internal user. The client has the impression of being connected to the real server whereas he/she is actually connected to an auxiliary server. The auxiliary server communicates securely with another remote auxiliary server in the LAN where the real one is located.

In the case of a remote-user to LAN communication, there is no server process at the user's end. It is the VPN-Insel client software that directly manages secure communications. Therefore, the Internet Service Provider, located in an intermediate point, need not be trustworthy or secure.

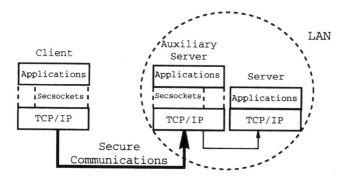

Fig. 10. Remote-user to LAN communication under VPN-Insel

In both cases, the VPN begins to work when the private network administrator starts the main server: the process *S1*. Then, the following events occur:

1. *S1* initializes the secondary servers database and the remote users database.
2. *S1* starts a secondary server for its own LAN (the LAN of the main entity).
3. *S1* remains open waiting for requests. The requests come from other LAN's secondary servers or from remote users.
4. Each secondary server, started by the corresponding LAN administrator, is registered in the main server, in order to integrate the LAN into the VPN.
5. Each secondary server stays open for: a) requests from users in its own LAN; or b) requests from remote auxiliary servers that want to use its LAN's resources.
6. When a client requests a communication to its secondary server (or, depending on the case, to the main server), this one creates an auxiliary server that exclusively manages that secure connection. Afterwards, the secondary server (or the main one) remains open for new requests.

Our VPN solution would not be complete if it was necessary to modify the software of other traditional network applications previously installed in the computer systems of the organization branch offices. The next section shows how we avoid modifications to previously installed software in a simple way. In this case, the example refers to an FTP service, although it is similar for other services. The section also describes, for such service, the interaction of the processes that VPN-Insel creates in a typical scenario for this kind of communication.

6   Operation Example and Implementation Issues

Generally, a user runs a client program in order to use an Internet service. Afterwards, the client program connects to the server program in the remote computer system. But in our solution the client does not run the client program. Instead, the user decides, by using the application interface of VPN-Insel, which network application to use, e.g., FTP. Then, making use of the same interface, the user introduces the name of the computer (or IP address) where the FTP server is running. At this moment, VPN-Insel runs the FTP client program, but it does not provide it with the computer address where the real FTP server is running. VPN-Insel provides the FTP Client with the address where the auxiliary server is running inside its own client LAN.

Normally, the client process would connect to a predetermined port (port #21 for FTP service). But, in this case, it will not find a FTP server there; it will find the auxiliary server, *Saux-C*, of the client LAN, that has been previously created by the secondary server where the FTP client has been redirected. In order to achieve it, ports to be used must be unused. For FTP case, the line *"ftp stream tcp nowait root /usr/sbin/in.ftpd in.ftpd"* of FTP service in the file */etc/inetd.conf* has to be deleted by the system administrator. Additionally, it is necessary to delete the line *"secftp 21/tcp"* from the file */etc/services*.

In this way, the FTP client connects to the auxiliary server which attends to the client, just like if it was the real FTP server. In fact, the auxiliary server will work as an intermediary in the communication.

Later on, as was shown in figure 11, the auxiliary server of the client LAN connects to the corresponding auxiliary server of the remote LAN. This auxiliary process is the one that really connects to the FTP server. VPN-Insel securely manages the connection between the two auxiliary servers by making use of the Secsockets interface. The figure shows the interaction of the processes created by VPN-Insel in the FTP communication scenario.

Following the action of the user throughout the application interface, the following sequence of events is set into action among the processes:

1. *C1*, started by the user, requests a connection to *S2*, its secondary server.
2. If the user is a legitimate VPN user, *S2* creates an auxiliary server, *Saux-C*, to manage the connection.
3. At the same time, process *C1* runs the FTP client software.
4. FTP client software connects to *Saux-C*.

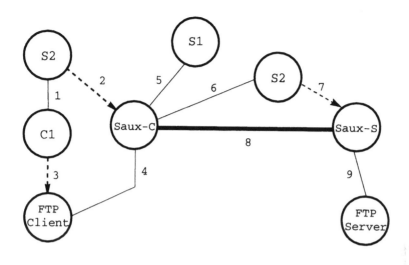

Fig. 11. Communication scenario for FTP service

The local auxiliary server, *Saux-C*, must connect to a remote auxiliary server, *Saux-S*, that guarantees secure communications with the node where the FTP server is located. *Saux-C* will request the secondary server *S2'* to create *Saux-S*. In order to achieve this, *Saux-C* has to know how to allocate *S2'*.

5. *Saux-C* requests *S1* for the number of the port and the host where process *S2'* is running.
6. *Saux-C* requests *S2'* to create *Saux-S*, the remote auxiliary server.
7. *S2'* creates *Saux-S*.
8. *Saux-C* connects to *Saux-S*.
9. *Saux-S* connects to the FTP server.

As shown in the previous figure, secure communication occurs between the auxiliary servers (thick line). They work as routers, receiving the packets, decrypting the information, and re-sending it to the final destination.

7   Conclusions

The study of the different TCP/IP stack layers reveals that the different solutions that enable establishing a VPN pay most of their attention to security aspects. These solutions focus on the isolation of a distributed network from outsiders and on the protection of sensitive information traversing Internet. But the users of these systems cannot freely use traditional services such as electronic mail exchange with non-VPN users, and cannot freely access Web and FTP servers external to the organization

In this paper we have presented a new solution for the implementation of VPNs at the transport layer that, while maintaining strong security features, allows the open use of traditional Internet services. As a consequence, it enables access to any remote node in the VPN and outside the VPN. The solution is exclusively software, so its deployment does not require the addition of new hardware or the modification of any existing one.

The solution allows an homogeneous control of all network services, which is necessary to gain higher user confidence regarding the entire system. Moreover, it does not require the modification of the software of traditionally insecure applications such as FTP, HTTP, Telnet, electronic mail, etc. As a result, the scheme is a simple and cheap solution for those organizations that want to install a VPN.

References

1. R. Atkinson, "Security Architecture for the Internet Protocol", RFC 1825, August 1995.
2. U. Black, "Frame-Relay: Specifications and Implementations", McGraw-Hill, 1994
3. CCITT, Recommendation X.509. *The Directory-Authentication Framework*. Blue Book - Melbourne 1988, Fascicle VIII.8
4. T. Dierks, C. Allern, "The TLS Protocol Version 1.0." Internet Draft, November 1998
5. W. Diffie, M. Hellman, "New Directions in Cryptography". *IEEE Transactions on Information Theory*, IT-22, n. 6. 1976, pp. 644-654.
6. FIPS 46, *Data Encryption Standard*, NBS, U.S. Department of Commerce, Washington D.C., January 1977
7. FIPS-140-1, *Security Requirements for Cryptographic Modules*, U.S. Department of Commerce, NIST, Washington, DC, 1994.
8. FIPS 180-1, *Secure Hash Standard*, NIST, U.S. Department of Commerce, Washington D.C., April 1995
9. R. Harbison, "Frame-Relay: Technology for our Time", *LAN Technology*, December 1992
10. M. Horowitz, S. Lunt, "FTP Security Extensions", RFC 2228, October 1997.
11. X. Lai, J. Massey, "Hash Functions Based on Block Ciphers" *Advances in Cryptology, Proceedings EUROCRYPT '92*, Springer-Verlag, 1992, pp. 55-70
12. J. Linn, "Privacy Enhancement for Internet Electronic Mail: Part I - Message Encipherment and Authentication Procedures", RFC 989, February 1987.
13. Microsoft Corporation, "The Private Communication Technology", 1997.
    http://premium.microsoft.com/msdn/library/backgrnd/html/msdn_pct.htm
14. Netscape Communications, "SSL 3.0 Specification".
    http://www.netscape.com/libr/ssl3/index.html
15. B. Ramsdell, "S/MIME Version 3 Message Specification", Internet Draft, August 1998.
16. R. Rivest, "The MD5 Message Digest Algorithm". RFC 1321, April 1992
17. A. Schiffman, E. Rescorla, "The Secure Hypertext Transfer Protocol", Internet Draft, June 1998.
18. B. Schneier, "Description of a New Variable-Lenght Key, 64-Bit Block Cipher (Blowfish)", *Fast Software Encryption*, Springer-Verlag, 1994, pp. 191-204
19. T. J. Socolofsky, C. Kale, "A TCP/IP Tutorial", RFC 1180, January 1991.
20. P.R. Zimmermann, "The Official PGP User's Guide". MIT Press, 1995.

Performance Evaluation of Certificate Revocation Using k-Valued Hash Tree

Hiroaki Kikuchi, Kensuke Abe, and Shohachiro Nakanishi

Dept. of Electrical Engineering, Tokai University,
1117 Kitakaname, Hiratsuka, Kanagawa, 259-1292, Japan
kikn@ep.u-tokai.ac.jp kensuke@ep.u-tokai.ac.jp

Abstract. A CRL (Certificate Revocation List) defined in X.509 is currently used for certificate revocation. There are some issues of CRL including a high communication cost and a low latency for update. To solve the issues, there are many proposals including CRT (Certificate Revocation Tree), Authenticated Dictionary, and Delta List. In this paper, we study CRT using k-valued hash tree. To estimate the optimal value of k, we examine the overhead of computation and the communication cost. We also discuss when a CRT should be reduced by eliminating unnecessary entries that are already expired.

1 Introduction

CRL (Certificate Revocation List) is the most common way to revoke certificate in public-key infrastructure (PKI) [1,2]. The CRL, however, has two problems; timeliness and communication cost [5]. The followings were proposed so far to solve the problems.

1. Certificate Revocation Tree (CRT) was proposed by Kocher [8]. By using hash tree instead of the list of CRL, a communication cost is decreased by $\log n$.
2. Authenticated Dictionary (AD) was proposed by Naor and Nissim [10]. A generalized hash tree so called "Authenticated Dictionary" is used not only for proving certificate revocation but also for updating database at directory.
3. Delta List (DL) was proposed by Kikuchi and Abe [11,12]. An implicit rule for tree updating ensures a uniqueness of CRTs at both CA and directory. The communication cost of "Delta List" method is smaller than that of AD. Whenever CRT is updated, a tree is balanced so that it minimizes average depth to the leaf and reduces a communication cost.

The difference between the AD and the Delta List is illustrated in Table 1. We developed an experimental implementation of Delta List, and showed the performance of proposed method in terms of computational and communication costs [11].

It's difficult to figure out the best revocation method because many factors are involved in. For example, when a revocation tree is balanced every time a

M. Mambo, Y. Zheng (Eds.): ISW'99, LNCS 1729, pp. 103–117, 1999.
© Springer-Verlag Berlin Heidelberg 1999

Table 1. The Difference between the AD and the DL

	Authenticated Dictionary	Delta List
What CA sends directory when it updates.	A partial tree which contains newly revoked certificates.	The list of the newly revoked certificates.
Communication cost	$O(\log_2 n)$	$O(1)$
Data structure	Random built tree	Balanced tree

certificate is revoked, the hash tree becomes surely shallow and the communication cost between directory and users is minimized. However, balancing may take too much time when certificates are revoked so often. Clearly, there is a trade-off between balanced and not balanced tree. The trade-off is driven by the ratio of certificate update rate to the verification rate.

The other factor concerns certificate expirations. An entry of expired certificate must be eliminated from the revocation list. Thus, any evaluation should take into account the outcome of the expiration processes.

In this paper, we extend the binary hash tree to the k-valued hash tree [14], and estimate performance of our the proposed method. The number of branches at a node, called "degree" of a tree, is monotonically increasing with k, while an average depth decreases with k. The hash computations are involved as many times as the degree of tree. But, a communication cost depends depth of a tree. In addition, the processing time on hash becomes small as k increases. Thus, it's not trivial to find the optimal k to minimize both costs. Hence, we present a process in which optimal degree of a tree can be identified given a environmental characteristics. Our evaluation is analytical supported by experiment.

2   Preliminary

2.1   PKI Model

In this paper, a PKI model consists of CA, a directory (server), and end users. A directory server provides a verification service by using the individual private key. The corresponding certificates are assumed to be widely distributed among users.

2.2   Certificate Revocation List (CRL)

A CRL is a list of revoked certificates and digitally signed by the CA. In X.509 framework [1], on a request from a user who wants to see if a certificate is already revoked or not, the directory server responds with the current entire CRL.

2.3   Online Certificate Status Protocol (OCSP)

An OCSP provides users more timely revocation information than what CRLs provide [5]. An OCSP response message consists of the response type and the

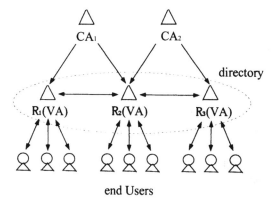

Fig. 1. PKI Model (Certification Authorities, directory servers and users).

revocation information that the particular certificate specified by the OCSP request. All response messages (including error messages) must be digitally signed by the private key of responder (a directory server) delegated by the CA [6,7]. At the expense of computational overhead for the digital signature, the OCSP enables timely checking with very low communication cost.

2.4   Certificate Revocation Tree (CRT)

A CRT is a digitally signed hash tree in which leaves represent revoked certificates identified by serial numbers, $X_1, \ldots, X_n$, where $X_1 < \cdots < X_n$. In a CRT, nodes are hash values computed for a concatenation of the child nodes. Let us suppose that a CRT has four leaves in Figure 2. A node, $X_{i,i+1}$, is given by

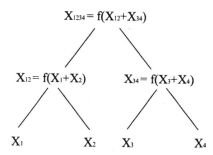

Fig. 2. Certificate Revocation Tree. Four certificates identified by serial numbers, $X_1, \ldots, X_4$ are jointly fed into an one-way hash function f.

$$X_{i,i+1} = f(X_i + X_{i+1})$$

where $f()$ is a collision intractable hash function and symbol $+$ denotes concatenation. In practical, any one-way hash function such as MD2, MD5[17] or SHA[18] can be used as a collision intractable hash function. In this example, the root hash value, $X_{root} = f((f(X_1) + f(X_2)) + (f(X_3) + f(X_4)))$, is dependent on all leaves and can not be forged by a new tree such that $X_{root} = X'_{root}$ and X'_{root} consists of other serial numbers than $X_1, \ldots, X_4$ under the assumption of collision intractable hash function $f()$.

On a verification request from user, a directory responds with a subtree which contains a path from a root to an appropriate leaf. When the requested certificate has already been revoked, a directory extracts the subtree consisting of the revoked certificate x_4 as shown in Figure 3.

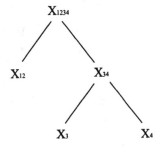

Fig. 3. The subtree to revoke x_4. The hash value X_{12} is used instead of X_1 and X_2.

3 CRT Update

3.1 Random Insertion

A naive CRT updating might result in having directory's CRT inconsistent with the CA's CRT. For instance, let us suppose that a newly revoked certificate, X_5, satisfies $X_3 < X_5 < X_4$. As shown in Figure 4 (a) and (b), there are two possible valid CRTs. In [10], the CA sends a subtree so that the directory is add to identify a node on which the newly revoked certificate is added. We ensure the uniqueness for updating CRT by having the following simple update rule:

Rule 1:

- Let X_i and X_j be the serial number of a newly revoked certificate, X, and the adjacent serial numbers of already revoked certificate, respectively. There is no revoked certificate X' such that $X_i < X' < X$ or $X < X' < X_j$.
- Let d_i and d_j be the *depth* of X_i and X_j, respectively. A depth is the length of a path from the root to the specific node.

Table 2. List of symbols

n	a number of certificates.
s	a number of revoked certificates.
k	a number of branches from a node.
d	an average depth of tree.
L	a size of partial tree.
B	a bandwidth of network.
T_r	a processing time to add entry into a tree.
T_b	a processing time to balanced tree.
L_r	a size of partial tree in randomly built tree.
L_b	a size of partial tree in balanced tree.
u	a CRT update rate.
v	a verification requested rate.
c_k	a number of hash computations in a partial tree.
C_k	a number of hash computations in the entire tree.
T_{hash}	a computational time for k-valued hash.
T_{at}	a turn around time of verification request.

- If $d_i < d_j$, insert X at the right side of X_i (see Figure 4 (a)). Otherwise, insert at the left side of X_j (Figure 4 (b)).
- If $d_i = d_j$, insert X at the right side of X_i.

Under the rule 1, a directory can uniquely figure out at which nodes the newly revoked certificate will be inserted. What the CA has to send out is the serial numbers of the revoked certificate and the digital signature, $\sigma(X_{root})$, computed with the root hash of the updated tree. The rule 1 reduces a communication cost between CA and directory up to a constant value necessary for specifying newly revoked certificates.

3.2   Balancing Update

We assume that a revocation of certificates happens independently. According to rule 1, a CRT may grow to a tree which has nodes of various depth, called a *randomly built tree*. It has known that a expected value of mean depth of a randomly built binary tree with n nodes, a_n, is given by

$$a_n = 2(H_n(n+1)/n - 2)$$

where H_n is a harmonic number, defined by $H_n = 1 + 1/2 + 1/3 + \cdots$ [14].
   A randomly built tree can be dealt with by small overhead for updating, but requires the communication cost proportional to the average depth. On the other hand, a *completely balanced tree*, which is a tree such that at most one difference in depth exists in left and right subtrees for each node, will minimize the communication cost by $\log_2 n - 1$ in the cost for updating the entire tree.

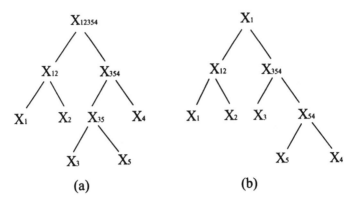

(a)   (b)

Fig. 4. Uniqueness. There are two possibilities to add a new entry X_5 into the tree in Figure 2.

Given any n, the tree can be completely balanced by the following steps:

---

btree()
Input: Let $X_1, \ldots, X_n$ be the serial numbers of revoked certificates in order.
Step 1: If $n \leq 2$, return tree of (X_1, X_2).
Step 2: Otherwise, compute $Y_1 = \text{btree}(X_1, \ldots, X_{\lfloor n/2 \rfloor})$
    and $Y_2 = \text{btree}(X_{\lfloor n/2 \rfloor + 1}, \ldots, X_n)$, and return (Y_1, Y_2).
Output: An S-expression corresponding to the desired completely balanced tree.

For instance, a set of certificates with $n = 9$ will be balanced as $(((X_1 X_2)(X_3 X_4))((X_5 X_6)((X_7 X_8)X_9)))$, which expresses a tree structure of Figure 5.

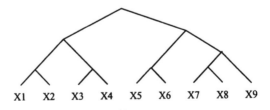

Fig. 5. Sample of completely balanced tree. Average depth from a root to leaves is guaranteed to be minimal.

4   The Update Problem of CRT

4.1   When Expired Certificate Should Be Deleted from CRT

All certificates revoked before expiration for some reason are listed in CRT. From the security viewpoint, there is no problem even if expired certificates are still in CRT. But, in terms of bandwidth spent by CRT transmission, we want keep CRT as small as possible, though too frequent deleting spends much bandwidth between CA and directory. So, we consider a problem when expired certificates are deleted from CRT. Generally, we have two options;

1. Directory update CRT independently from CA's update (Independent update).
2. Directory update CRT only when CA tells it (Sequential update).

The first protocol (independent update) is more efficient than the latter one (sequential update) because no communication between CA and the directory happens to update CRTs. But, the security highly depends on an assumption that clock are synchronized between CA and the directory. For example, let us suppose that the time at the directory is somehow proceeded. The directory deletes revoked certificate from CRT earlier than its true expiration date, and then it cannot prevent the stolen certificate from being abused. There is no such problem of security in the latter protocol (sequential update). However, it's a less efficient that the CA sends to the directory a message every time a certificate in CRT is expired. Therefore, we propose the next method (Rule 2).

Rule 2:

- When a newly revoked certificate, X_i, is added to CRT, CA sends the serial number, the expiration, E_i, and the signature on root hash $\sigma(Xroot)$ of root (same as Rule 1). The day of issue is added.
- On received information a directory updates the CRT. At the same time, it removes certificates X_j such that $t > E_j$, which have been expired before t.
- The directory computes the root hash of CRT, and verifies the signature.

The proposed protocol (Rule 2) is as secure as protocol 2 (sequential update) and as efficient as protocol 1 (independent update). Note that the amount of communication involved in this protocol is equal to that of protocol 1.

In this rule, certificates will not expire until some other certificate gets revoked, but it's not crucial to total security.

4.2   The Comparison between the Balanced Tree and Randomly Built Tree

Which is better, CRT is balanced, or not? We answer to the question by estimating the amount of processing in the directory.

Assumptions In our analysis, we make the following assumptions about the performance of CRT management system.

1. Processing time to cope with random tree or balanced tree is proportional to number of certificate in the directory. It's reasonable assumption because there are some well known data structures, such as linked list, which requires a linear overhead to add or delete on entry.
2. Communication cost is a size of average partial tree that might be sent over network multiple times on every transition.
3. Total performance of directory depends on how often CRT is updated and how often certificates are verified.

Processing Time Let T_r be a processing time that a newly revoked certificate is added to a given tree (a random tree), and T_b be a processing time to balance the tree. T_r and T_b are proportional to number of certificate, s. We assume that $T_r < T_b$. The processing time on SunUltra (S7/300U) 167MHz, Solaris2.5.1 is shown in Figure 6. Our implementation uses C, Perl, and sed.

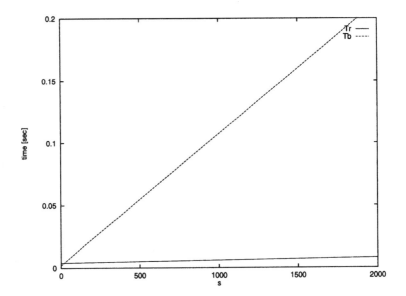

Fig. 6. Processing time to update CRTs. T_r is a time of randomly built tree; T_b is a time to balance tree. Both costs are linear to s (number of certificates in a directory).

Communication Cost The communication costs with respects to a number of certificate, s, are shown in Figure 7. Where, L_r and L_b are the sizes of partial hash tree, which are transmitted from the directory to the user, in the randomly built tree and a balanced tree, respectively. We see both increase by $O(\log s)$.

When a user wants to see the availability of a certificate, the user asks the directory to send a "partial hash" that consists of all hash values in a path from the root to particular leaf. Since depths in random tree are not uniform, we use the average depth for all leaves. We see that a random tree grows 1.31 times in size as deep as a balanced tree does in average.

In [14], it's shown that the expected depth of randomly built tree, $E[L_r]$, is approximated by

$$\frac{E[L_r]}{E[L_b]} = 2\ln(2) = 1.38,$$

where $E[L_b] = \lceil \log_k(s) \rceil$. Our estimate agrees with this analysis.

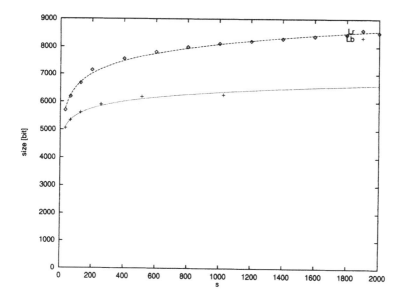

Fig. 7. Communication cost (Size of hash tree). L_r and L_b are sizes of partial hash tree in randomly built tree and balanced tree, respectively. Actual measured values are indicated by dots and crosses. Using the error minimizing algorithm, the fitting functions are displayed.

Relations between the Processing Time and the Communication Cost
We learn that there is a trade off between the processing time and the communication cost. So, we consider both conditions and estimate the overall processing time. The overall processing time of the randomly built tree, P_r, and a balanced tree, P_b, are given as follows;

$$P_r = uT_r + vL_r/B,$$

$$P_b = uT_b + vL_b/B,$$

where, u is a CRT update rate, that is, an average update number of times per a unit time, v is a verification requested rate from users, and B is a communication bandwidth. By having $P_r = P_b$, we have the following condition when a randomly built tree takes a time as much as a balanced tree,

$$uT_r + vL_r/B = uT_b + vL_b/B,$$

that is,

$$v/u = \frac{T_b - T_r}{L_r - L_b} B. \tag{1}$$

Where, v/u is the ratio of the update rate to the verification rate. When v/u is small, a random tree is better. Fitting data of Figure 6 and Figure 7 with a function of s gives Figure 8, in which the boundary of randomly built tree and balanced tree are indicated. We see that v/u increases monotonously with the number of revoked certificate, s. Therefore, a randomly built tree is better when s is large. In a standard PKI environment, $v/u = 100$ is said to be a reasonable value [19].

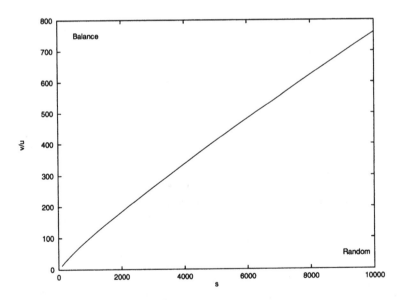

Fig. 8. Relations between u and v. u and v are update and verification rates. Boundary function divides a parameter space into two areas; in the left and top area, the balanced approach is better and the right and bottom shows area where the randomly built tree is appropriate.

5 k-Valued Hash Tree

In this section, we expand CRT to the k-valued hash tree, and discuss the optimal k in terms of processing time and the communication cost. See sample trees in Figure 10. Throughout this section, we don't care about the difference between random and balanced tree because we want to focus on the communication cost between directory and users. We assume balanced tree when necessary.

5.1   Communication Cost

First of all, we examine a communication cost between the directory and the user. A communication cost depends on a size of a partial tree which is of a path from a root to any leaf in the given tree. Clearly, as k increase, the depth of tree decrease, and the communication cost seems to decrease correspondingly. However, it should be noticed that a number of hash values to be sent increase at the same time. For example, we need sending two hash values, X_{123} and X_{456}, to prove X_7 belonging to tree (a) in Figure 10. When $k = 4$ in tree (b), three hash values, X_{1234}, X_{6789} and $X_{10111213}$, are required to prove X_5.

More generally, a number of hash values in k-balanced partial tree is given by $c_k = d(k - 1)$, where, d is the average depth computed by $d = \lceil \log_k(s) \rceil$. Hence, we have a size of partial tree L as follows:

$$L = \alpha c_k + \beta = \alpha(k - 1)\lceil \log_k(s) \rceil + \beta, \tag{2}$$

where α is a bit length per one hash, and β is a constant that doesn't depend on s. According to the sample CRL generated by SSleay0.9b, we assign the size of CRL as α=64[bit] and β=3608[bit], and demonstrate the performance in Figure 9. The figure shows that size L increases as s increases, and similarly as k increase. Therefore, we see that $k = 2$ minimizes the communication cost for s.

5.2   The Processing Time to Compute Partial Trees

Second, we estimate processing time to compute k-valued partial trees. A processing time is proportional the number of hash computations in the entire tree. Suppose that CRT is completely balanced. A number of hashes in a binary tree, C_2, is $C_2 = s - 1$.

More generally, we have a number of hash computations performed times in a k-valued tree, as

$$C_k = \lceil \frac{s - 1}{k - 1} \rceil. \tag{3}$$

Tree (a) is an example of 3-valued tree, which has a total of three hash computations necessary to compute as equation (3) gives. Indeed, Tree (a) has three nodes of X_{123}, X_{456}, and X_{root}. As well, 4-valued tree (b) has a total of four nodes, which agrees that $C_4 = (13 - 1)/(4 - 1) = 4$. We show the relation

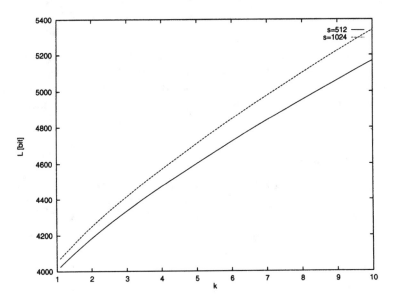

Fig. 9. A size of a partial tree with k-valued hash with respects to k. The behavior of L represents the expected communication cost.

between k and C_k in Figure 11. The number of hash computations decreases as k increases. Hence, we see $k = s$ minimizes the processing time at a directory. Note that this is the case of CRL.

5.3   The Optimal Value of k for Both the Processing Time and the Communication Cost

As we have seen, there is a "trade-off" between a hash times C_k and a size L. To take both behaviors into account, we estimate an average turn around time of verification request from users. By letting T_1 be a time to compute one hash, we have a computational time for k-valued hash as

$$T_{hash} = C_k T_1 = \lceil \frac{s-1}{k-1} \rceil T_1.$$

A time to transmit an L-bit request from a user to a directory is

$$T_{ex} = L/B$$

where B is a network bandwidth between the directory and the user. Finally, we have the total estimate for turn around time, T_{at}, as follows:

$$T_{at} = C_k T_1 + L/B = \lceil \frac{s-1}{k-1} \rceil T_1 + \frac{\alpha(k-1)\lceil log_k(s) \rceil + \beta}{B}. \qquad (4)$$

We demonstrate equation (4) in Figure 12 with parameters T_1=0.677[ms], and B=1.5[Mbps]. In this particular example, k=85 is the optimal value for s=512,

(a)

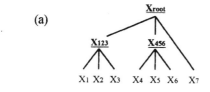

(b)

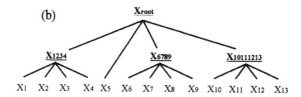

Fig. 10. Example of 3-valued and 4-valued hash trees.

and $k=120$ for $s=1024$. Please note that in this evaluation the hash computations are done every time. With caching intermediate results, the optimal value of k would be smaller. Also note that on optimal value depends a given environment including system parameters, which maybe dynamic, unknown or unmeasurable in some cases.

6 Conclusion

We estimated the effect of balancing in CRT tree in consideration of the update frequency. We extended a binary tree in the k-valued tree, and analized the performance in terms of degree.

The main results are as follows;

1. There is a trade-off between random and balanced certificate revocation trees. The trade-off is driven by the ratio of certificate verification to revocation.
2. The degree of revocation tree has an impact on both tree update and user-directory communication costs. We show a process in which approximately optimal degree of tree can be identified given environmental characteristics.

Acknowledgment

The authors thank anonymous reviewer for his valuable comments and corrections.

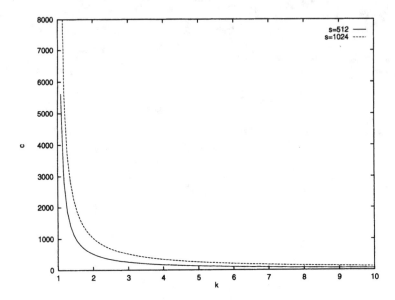

Fig. 11. Number of hash computations performed in a k-valued tree. This means the average computational overhead at directory.

References

1. ITU-T Recommendation X.509—ISO/IEC 9594-8: 1995
2. R. Housley, W. Ford, W. Polk, D.Solo, Internet X.509 Public Key Infrastructure Certificate and CRL Profile, Internet RFC 2459, 1999
3. C. Adams, S. Farrell, Internet X.509 Public Key Infrastructure Certificate Management Protocols, Internet RFC 2510, 1999
4. S. Boeyen, T. Howes, P. Richard, Internet X.509 Public Key Infrastructure Operational Protocols - LDAPv2, Internet RFC 2559, 1999
5. M. Myers, R. Ankney, A. Malpani, S. Galperin, C. Adams, X.509 Internet Public Key Infrastructure Online Certificate Status Protocol - OCSP, Internet RFC 2560, 1999
6. R. Housley, P. Hoffman, Internet X.509 Public Key Infrastructure Operational Protocols, Internet RFC 2585, 1999
7. S. Boeyen, T. Howes, P. Richard, Internet X.509 Public Key Infrastructure LDAPv2 Schema, Internet RFC 2587, 1999
8. P. Kocher, A Quick Introduction to Certificate Revocation Trees (CRTs), http://www.valicert.com/company/crt.html
9. R. C. Merkle, "A certified Digital Signature", Proc. Crypto '89, Lecture Notes in Computer Science 435, pp. 234-246, SpringerVerlag, 1989
10. Mni Naor and Kobbi Nissim, Certificate Revocation and Certificate Update, in proc. of Seventh USENIX Security Symposium, pp.217-228, 1998
11. H. Kikuchi, K. Abe and S. Nakanishi, Proposal on Update of Certificate Revocation Tree Using k-valued Hash Tree (in Japanese), Proc. of the 1999 Symposium on Cryptography and Information Security (SCIS'99), Vol.2, pp.621-626, 1999

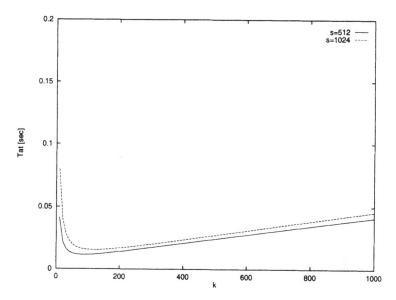

Fig. 12. Turn around time (T_{at}) with respects to k. Turn around time shows a total performance of directory. Given a number of certificate, s, there is optimal value of k.

12. H. Kikuchi, K. Abe, S. Nakanishi, Performance Evaluation of Public-key Certificate Revocation System with Balanced Hash Tree, proc. of International Workshop on Security (IWSEC'99), 1999 (to be appeared)
13. H. Kikuchi, K. Abe, S. Nakanishi, Certificate Revocation and Update Using Binary Hash Tree (in Japanese), IPSJ Technical Report, 98-CSEC-2, pp.51-56, 1998
14. N. Wirth, *ALGORITHMS AND DATA STRUCTURES*, 1986
15. E. Yang, SSLeay, http://www.ssleay.org
16. R. Rivest, "The MD4 message-digest algorithm", Internet RFC 1320, 1992
17. R. Rivest, "The MD5 message-digest algorithm", Internet RFC 1321, 1992
18. U.S. National Institute of Standards and Technology, "Secure Hash Standard", Federal Information Processing Standards Publication 180, 1993
19. H. kikuchi, et.al., An implementation and evaluation of certificates distributing system in privacy enhanced mail, IPSJ Trans., vol.36, No.8, pp.2063-2079, 1995

Active Rebooting Method for Proactivized System: How to Enhance the Security against Latent Virus Attacks⋆

Yuji Watanabe and Hideki Imai

Institute of Industrial Science
University of Tokyo
7-22-1 Roppongi, Minatoku, Tokyo 106-8558, Japan
mue@imailab.iis.u-tokyo.ac.jp, imai@iis.u-tokyo.ac.jp

Abstract. The notion of proactive security of basic primitives and cryptosystems was introduced in order to tolerate a very strong "mobile adversary[1][2][3][4]". However, even though proactive maintenance is employed, it is a hard problem to detect the viruses which are skillfully developed and latent in the memory of servers.

We introduce a new type of virus attacks, called *latent virus attack*, in which viruses reside in the intruded server and wait for the chance for viruses colluding with each other to intrude more than the threshold of servers.

The main subject of this paper is to analyze the resilience of proactive system against latent virus attacks and present how to enhance the security against such virus attacks.

At first, we estimate the robustness of proactivized systems against this attack by probabilistic analysis. As a result, we show that if the virus detection rate is higher than a certain threshold, it is possible for proactive maintenance to make the system robust, while, if less than the threshold, the failure probability of the system is dependent only on the virus infection rate.

In order to enhance the resilience against such virus attacks, we propose the notion of *active rebooting*, in which the system performs the reboot procedure on a predetermined number of servers in the total independence of servers being infected or not. We estimate the security of proactive maintenance with active rebooting by extending the probabilistic model of proactive maintenance. As a result, we show that active rebooting enables us not only to enhance the security against the viruses with higher infection rate, but also to make the system robust even in the case of a low detection rate. Moreover, we show that it is effective even in the case the number of servers which are forced to carry out the reboot operation every update phase is comparatively small.

⋆ This work was performed in part of Research for the Future Program (RFTF) supported by Japan Society for the Promotion of Science (JSPS) under contact no. JSPS-RFTF 96P00604.

M. Mambo, Y. Zheng (Eds.): ISW'99, LNCS 1729, pp. 118–135, 1999.
© Springer-Verlag Berlin Heidelberg 1999

1   Introduction

Threshold protocols[5][6][7] address a variety of adversaries and a variety of attacks. They maintain appropriate security against illegal hackers, insiders, disgruntled ex-employees, computer viruses, and other agents of data espionage and destruction.

Threshold protocols maintain secrecy in the face of up to $k - 1$ adversaries and yet achieve data integrity and availability with the cooperation of k out of n shareholders. They are called "(k, n) *threshold protocol,*" which is based on the presupposition that the ability of attacker is less than a predetermined threshold. There are many researches on upper bounds of threshold in different models, and various solutions have been proposed in order to achieve these upper bounds efficiently.

However, it is substantially difficult to estimate the ability of adversary quantitatively and to set reasonable threshold, especially in case they are an attractive target for break-ins. For instance, we consider a $(3, 5)$ threshold protocol (in practice, MasterCard/Visa SET root key certification system has been implemented as a 3 out of 5 RSA threshold signature scheme[8].) This protocol assures the security in a cryptographic sense, provided the adversary can corrupt (during the entire lifetime of the protocol) only at most 2 servers. It is natural that we should concern what the vulnerability of the system is, in other words, "what is the probability that the adversary can corrupt more than 3 servers and the assumption of the protocol is exploded?"

Moreover, some system, such as certification authorities (CA), must remain secure for very long period of time. In these systems, it is intuitively clear that it is difficult to restrict the attack within a constant fraction of servers during an entire lifetime of the system. (Given a sufficient amount of time, an adversary can break into servers one by one, thus eventually compromising the security of the system.)

A trivial but effective method to enhance the security against this fault is to set the threshold high enough to prevent the attack. Upper bounds of possible thresholds are therefore to be chosen as the proportion to the number of whole servers (typically the majority of servers), i.e., this method results in increasing the number of participating servers. Therefore, disadvantages of this approach include the complicated management of server's security.

Recently, the notion of "proactive security[1]" of basic primitives and cryptosystems that are distributed amongst servers was introduced in order to tolerate a very strong "mobile adversary" without increasing the number of servers. This adversary may corrupt all servers throughout the lifetime of the system in a non-monotonic fashion (i.e. recoveries are possible) but the adversary is unable to compromise the secret if at any time period it does not break into more than $k - 1$ locations. ($k = \lfloor l/2 \rfloor$ is optimum[1].) This setting was originally presented by Ostrovsky and Yung[1]. The notion of "proactive security" assures increased security and availability of the cryptographic primitive [2][3][4].

Proactive maintenance adds a periodic refreshing of the contents of the distributed servers' memories. Therefore, the knowledge of the mobile adversary (representing: hackers, viruses, bad administrator, etc.) obtained in the past from compromising at most the allowed $k - 1$ number of servers is rendered useless for the future. As a result, the system can tolerate a "mobile adversary" which is allowed to potentially move among servers over time with the limitation that it can only control up to $k - 1$ servers during a period of time[9].

1.1 Latent Virus Attack to Proactive System

We consider a very pragmatic scenario. Very diverse viruses are produced by malicious one and try various possible means to infect servers and to avoid the detection. Once the existence of a virus is detected by virus detection tools (e.g. anti-virus scanners) or checking protocols at run-time (e.g. VSS[10][11]), the system instantly triggers the reboot operation of the infected server in order to remove the virus from the server completely[1][2].

Computer viruses (as well as biological viruses) go through several processes (infection $\rightarrow$ latency $\rightarrow$ activation) until they cause damage to infected servers. Once the viruses become active and disturb the system in a malicious way, they are removable by checking the protocol or by detecting the infection. However, how can we detect and remove the virus in the latent period? Proactive maintenance provides the measure to find malicious behavior, but does not provide the measure to detect latency of virus. Therefore, detection of latent virus relies on virus detection tools. However, it is an intuitively hard problem to detect the viruses which are skillfully developed and latent in the memory of servers with bated breath. Cohen[12] showed that a perfect defense against computer viruses is impossible. In fact, we never know perfect virus detection tools which detect all latent viruses without exception before their activation.

Furthermore, we modify the model of latent virus to a more powerful one which adaptively causes malicious corruption. In this model, latent viruses reside in the intruded server with bated breath and wait for the chance for viruses colluding with each other to intrude more than the threshold of servers. Once the chance has come, all viruses become active and compromise the security of the system.

Of course, a latent virus setting seems to be rather theorical but must be a potential threat, because we can easily see its reality by observing a vicious spiral of construction of virus detection tools and appearance of new type of stronger virus. A latent infection is a potential threat over many network-systems, but is hardly manageable by cryptographic techniques.

1.2 Our Results

The notion of proactive security was introduced in order to realize the long-term security against break-ins. Proactivized system includes the mechanism to refresh the servers' memories and renew the exposed shares into new one in order to make the old one useless for the adversary, as well as the mechanism to detect malicious behavior during the protocol.

However, proactive systems provide no measure to detect the latent viruses before their activation, so it depends on the ability of the virus detection tools. Therefore, a proactive system is not robust against the attack by latent viruses, due to the hardness to detect latent viruses which is produced skillfully.

The main purpose of this study is to show a method for enhancing the robustness of proactive protocol against latent virus attacks. Our contribution in this paper is as follows.

Estimation of the Robustness of Proactive Security against Latent Virus Attacks At first, we examine the robustness of proactive system against the attack caused by latent viruses. Since latent viruses can be detected only by the virus detection tools, the robustness of the system is represented as a function which has the infection rate and detection rate as parameters[13]. Nevertheless, due to the difficulty to evaluate both parameters quantitatively, there have been few efforts to analyze virus infection and detection on proactivized servers theoretically. In fact, most of the discussions in this concern are mainly based on empirical knowledge.

In [12], Cohen pointed out that a perfect defense against computer viruses is impossible. Kephart and White[14] first studied the probabilistic model of virus infection by adapting mathematical epidemiology to this problem. They show an imperfect defense against computer viruses can still be highly effective in preventing their widespread proliferation, provided that the infection rate does not exceed a well-defined critical epidemic threshold. We modify their probabilistic approximation of virus infection and detection in order to make it suitable for the proactivized systems. Then, we present some tradeoff among an infection rate of viruses, a detection rate of servers and a failure probability of the system in which a certain number of servers are engaged in a proactive protocol. As a result, we show the fact that if the virus detection rate is higher than a certain threshold, it is possible for proactive maintenance to make the system robust, while, if less than the threshold, the failure probability of the system is dependent only on the virus infection rate.

Analysis of the Effect of Active Rebooting The inefficiency of proactive maintenance, in case servers can detect the infection of viruses at low rate, is due to the fact that the proactive maintenance does not work well for preventing viruses from being latent, as long as triggering the reboot operation and refreshing the memory is not until the viruses are detected (*passive rebooting*).

Passive rebooting is not effective in coping with latent virus attacks, as long as the virus detection rate is low. Accordingly, we propose the notion of *active rebooting*, in which the system performs the reboot procedure on a predetermined number of servers in the total independence of servers being infected or not. We estimate the security of proactive maintenance with active rebooting by extending the probabilistic model of proactive maintenance.

As a result, we show that active rebooting enables us not only to enhance the security against the viruses with higher infection rate, but also to make the system robust even in the case of a low detection rate. Moreover, we show that it is effective even in the case that the *level* of active rebooting (which means the number of servers which are forced to carry out the reboot operation in every update phase) is comparatively small.

1.3 Organization

The organization of this paper is as follows. In the next section, we explain the basic definitions and a probabilistic model of virus infection and detection in a proactive maintenance. In section 3, we estimates the robustness of a proactivized system against latent virus attacks. Then, as an effective and practical countermeasure against such attacks, we show "active rebooting" method in section 4. Finally, we shall conclude in section 5 with a summary of our results and future works in this area.

2 Preliminaries

2.1 Definition

We basically use the model of [2] with a slight modification. This modification is due to simplification of the analysis on latent virus attacks (See definition 4).

Definition 1 ((k, n)-threshold protocol[15]).   *A system of n servers $\mathcal{P} = \{P_1, \ldots, P_n\}$ share a function f_s for some key s. This secret value s is shared among n servers through a (k, n)-threshold scheme (i.e., s is divided into n shares $\{s_1, \ldots, s_n\}$ such that $k - 1$ shares provide no information on the secret, while k shares suffice for the reconstruction of the secret). We say that the protocol is a (k, n)-threshold protocol if, when k uncorrupted servers are active, for any x, the shared function f_s can be reconstructed to compute $f_s(x)$ even in the presence of $k - 1$ corrupted servers, yet nothing about f_s is revealed other than $f_s(x)$.*

Generally, for any f, this protocol is also called *threshold function sharing*[15]. Especially, if $f_s()$ outputs s itself(one time scheme), this protocol is called *threshold secret sharing scheme*[5].

Definition 2 ((k, n)-proactive protocol[2]). (k, n)-proactive protocol *is a variant of (k, n)-threshold protocol in order to enhance the security against the adversary who can corrupt all servers throughout the entire lifetime of the system in a non-monotonic fashion, but can corrupt no more than $k - 1$ out of n servers during any period of time. Proactive protocol works as follows*

1. *Let t be a time which have passed from the beginning of the system. For instance, we assume a unit time is a day. The lifetime of the secret s is divided into periods of time (e.g. a day, one week, etc.). We define ξ as $\xi = \lfloor t/\tau \rfloor$ where τ is the length of single time period. Accordingly, if $(\xi-1)\tau \leq t < \xi\tau$, we say that the system at the time t belongs in ξ-th time period.*

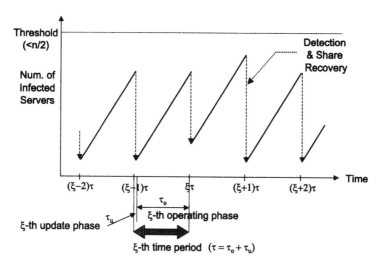

Fig. 1. A diagram illustrating proactive maintenance

2. *Each time period is divided into two parts,* operating phase *and* update phase. *The time length of them are denoted as* τ_o *and* τ_u, *respectively (i.e.,* $\tau = \tau_o + \tau_u$), *where the update phase is negligibly short when compared to the length of a time period, that is,* $\tau_u \ll \tau$, $\tau_o \approx \tau$. *We refer to the period within* $(\xi-1)\tau \le t < (\xi-1)\tau + \tau_u$ *as* ξ-th update phase $\mathcal{U}^{(\xi)}$, *and the period within* $(\xi-1)\tau + \tau_u \le t < \xi\tau$ *as* ξ-th operating phase $\mathcal{O}^{(\xi)}$. *The largest part of each time period belongs to the operating phase and only a short period at the beginning of each time period belongs to the update phase.*

3. *At update phases, the servers engage in an interactive update protocol, after which they hold completely new shares of the same secret. A non-faulty majority of servers trigger a reboot operation of faulty servers during an update phase, in order to bring a completely fresh version of the program from ROM to memory. We refer these operations during update phases as* proactive maintenance.

Note that a server which has been infected by the virus but has not detected it yet does not perform the reboot operation. Accordingly, the viruses can stay in servers' memories as long as it is not detected. Considering the behavior of such viruses is the main subject of this paper.

For the simplicity of analysis, we do not consider the corruption during an update phase. Even if a server is corrupted during an update phase, we can consider the servers as corrupted during both periods adjacent to that update phase. It is also not a realistic concern in our setting, where the update phase is negligibly short when compared to the length of a time period.

Removablity of Viruses through Reboot Procedure

We assume that the adversary intruding the servers $\mathcal{P}$ is "*removable*" in the sense that all of the shares are thrown away and new shares of the secret are created. These was carried out through a *reboot* procedure when it is detected by explicit mechanisms by which a majority of (honest) servers always detects and alerts about a misbehaving server or by regular detection mechanisms (e.g., anti-virus scanners) available to the system management.

Triggering the reboot operation of a misbehaving server relies on the system management which gets input from a majority of (honest) servers. Once the mechanism is found to be infected, a complete reboot is performed in order to bring a completely fresh version of the program from ROM to memory. In this paper, we assume that viruses cannot survive a physical reboot of the machine (for its detail, see [2]). and for this assumption, we also assume each server has the minimum amount of trusted hardware for I/O and ROM.

2.2   The Model of Virus Infection and Detection

One of the purposes of this paper is to estimate the robustness of the system based on a threshold structure against virus attacks.

Generally, at first, viruses attempt to intrude into server's memory. After they are latent within a certain period of time after their intrusion, they act and behave in various malicious ways. However, once such viruses act, they are detectable with extremely high probability by the virus detection mechanism of the system. Accordingly, the powerful attacker who attempts to corrupt the system totally design viruses which are adaptively latent until intruding more than a threshold of servers. Therefore, we consider adaptive viral behavior such that viruses does not act until they succeed in intruding and lurking into more than a threshold of servers.

According to these observations, the state of an individual server can be placed in the following three categories. Let $\mathcal{C}(t)$ be a set of servers which are uninfected by viruses at the time t, $\mathcal{I}(t)$ be a set of servers which are infected by viruses but not aware of their own infection, and $\mathcal{D}(t)$ be a set of servers which are infected by viruses and have already detected the infection of viruses in their own memories. Define $n_c(t) = |\mathcal{C}(t)|$, $n_i(t) = |\mathcal{I}(t)|$ and $n_d(t) = |\mathcal{D}(t)|$. Note that $n = n_c(t) + n_i(t) + n_d(t)$ (because $\mathcal{C}(t) \cap \mathcal{I}(t) = \phi$, $\mathcal{I}(t) \cap \mathcal{D}(t) = \phi$, $\mathcal{C}(t) \cap \mathcal{D}(t) = \phi$ and $\mathcal{C}(t) \cup \mathcal{I}(t) \cup \mathcal{D}(t) = U$.)

Figure 2 shows the simple model of state transition with respect to virus infection and detection. Let β, called *infection rate*, be the probability with which an individual uninfected server is infected by viruses in a unit time, i.e.,for $\forall \xi$, $\forall t \in \mathcal{O}^{(\xi)}$, $\forall P_{j'} \in \mathcal{C}(t)$,

$$\Pr[P_{j'} \in \mathcal{I}(t + \Delta t)] = \beta \Delta t \quad \Pr[P_{j'} \in \mathcal{C}(t + \Delta t)] = 1 - \beta \Delta t.$$

After a virus intruding and being latent in $P_{j'} \in \mathcal{C}(t)$, $P_{j'}$ belongs to $\mathcal{I}(t + \Delta t)$. Consequently, $P_{j'} \in \mathcal{I}(t + \Delta t)$, $n_c(t + \Delta t) = n_c(t) - 1$ and $n_i(t + \Delta t) = n_i(t) + 1$.

Proactivized servers should take enough measure to prevent a virus infection from propagating successively by strict management of communication among

Operating Phase

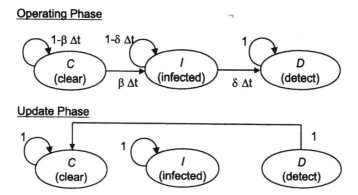

Update Phase

Fig. 2. A model of virus infection and detection on proactive system

servers or use of diverse operating systems among each other. If not, it may be feasible for the number of infected servers to exceed the predetermined threshold. So, we do not consider the successive propagation of virus infection but consider the simple model that the infection rate β is invariable throughout the lifetime of the system.

Moreover, most of proactivized system is symmetrically constructed, which means that there is no difference among the roles the different servers play. So, we use the model that each individual server is infected by viruses at the same rate β among each other.

Similarly, let δ, called *detection rate*, be the probability with which an individual infected server become aware of viruses intruding and being latent in its inside in a unit time, i.e., for $\forall \xi$, $\forall t \in \mathcal{O}^{(\xi)}$, $\forall P_{j'} \in \mathcal{I}(t)$,

$$\Pr[P_{j'} \in \mathcal{D}(t + \Delta t)] = \delta \Delta t \quad \Pr[P_{j'} \in \mathcal{I}(t + \Delta t)] = 1 - \delta \Delta t$$

After $P_{j'} \in \mathcal{I}(t)$ detected the infection of virus, $P_{j'}$ belongs to $\mathcal{D}(t + \Delta t)$. Consequently, $P_{j'} \in \mathcal{D}(t + \Delta t)$, $n_i(t + \Delta t) = n_i(t) - 1$ and $n_d(t + \Delta t) = n_d(t) + 1$. Viruses detected in $\mathcal{O}^{(\xi)}$ are surely removed from server's memory as a result that the reboot procedure is triggered with probability 1 in the update phase $\mathcal{U}^{(\xi+1)}$ (This setting is well defined in [2]).

At each update phase, the system performs proactive maintenance in order to renew old shares of all servers into new ones. The update phase is negligibly short when compared to the length of a time period, that is, $\tau_u \ll \tau$. Therefore, we assume that there is not any new infection and detection within an update phase, because it is reasonable that the probability of occurrence of new infection and detection within an update phase is negligibly small compared with that within an operation phase. Accordingly, infection and detection of viruses can be defined as

$$\forall \xi, \ \forall t \in \mathcal{U}^{(\xi)}, \ \forall P_{j'} \in \mathcal{C}(t), \quad \Pr[P_{j'} \in \mathcal{C}(t + \Delta t)] = 1$$

$$\forall \xi, \ \forall t \in \mathcal{U}^{(\xi)}, \ \forall P_{j'} \in \mathcal{I}(t), \quad \Pr[P_{j'} \in \mathcal{I}(t + \Delta t)] = 1$$

We consider in this paper the probability of occurrence of failure caused by latent viruses intruding into servers. Accordingly, we define the failure of (k, n)-threshold (proactive) protocol as follows.

Definition 3 (Failure of threshold protocol). *For a (k, n) threshold (proactive) protocol, we say that the system fails if the number of corrupted (or infected) servers exceeds $k - 1$ at some point of time t. $P_f(t)$ denote the probability of occurrence of this failure from the beginning of the system till the time t. We call $P_f(t)$ the failure probability, which can be defined as*

$$P_f(t) := \Pr[\exists \zeta < t, \ n_i(\zeta) + n_d(\zeta) \geq k].$$

2.3 Latent Virus Attack

At the end of this section, we informally define the adversary that we consider in this paper.

Definition 4 (Latent virus attack). *Let $\mathcal{P} = \{P_1, \ldots, P_n\}$ denote a system of n servers employing the (k, n)-proactive protocol. We call a group of adversary $\mathcal{A} = \{A_1, \ldots, A_n\}$ latent viruses if $\mathcal{A}$ meets the following properties.*

- *$\mathcal{A}$ is controlled by a malicious person, called director.*
- *A_j can infect server P_j in a unit time at the rate β.*
- *After the infection, A_j is latent in server's memory without carrying out malicious behavior and wait for the instruction from the director.*
- *Once more than a threshold of servers have been infected by a subset of $\mathcal{A}$, denoted as $\mathcal{A}'$, director orders $\mathcal{A}'$ to corrupt intruded servers simultaneously in arbitrary malicious way, which may lead to disrupt the proactive protocol.*
- *$\mathcal{A}$ is computationally bounded and therefore can not break any of the underlying cryptographic primitives used.*

Notice that the director can wait for the opportunity to disturb the servers, since he knows whether the intrusion of viruses into servers results in success or failure. It is intuitively reasonable that viruses for intruding into proactivized servers are more adaptive and intellectual than biological viruses epidemic or the normal computer viruses.

We will take up the $(4, 7)$−proactive(threshold) protocol as a typical example. Our greatest concern in this paper is to enhance the security against latent virus attack without modifying the existing system, so we does not leave increasing the number of servers out of consideration (Of cource, it is one of the most effective solutions. However, it is trivial fact that increasing of the number of participating servers results in making the system more robust against the failure.) Narrowing an argument down to the case of a certain number of servers (here we take up 7 servers' case), we evaluate its security against latent virus attacks and show how to enhance the security without increasing the number of participating servers.

Throughout the following analysis, we assume the length of the time period τ is a day. So, β and δ mean the probability of occurance of new infection and detection within a day, respectively.

3   Analysis

3.1   Robustness against Latent Virus Attacks (Threshold Protocol without Proactive Maintenance)

At first, we examine a failure probability of (k, n)-threshold protocol without proactive maintenance against latent virus attacks. In order to account for the relationship between obscure parameters, we use the probabilistic approximation of virus infection and detection (This description was introduced by Cohen[12].) Accordingly, we do not deal with states of servers at any point of time as deterministic states but as probabilistic states by representing the probability of transition from one state to another.

Note that there are no transition in the direction of decreasing the number of infected servers, because the information of shares which has been leaked out by viruses cannot be recovered without being refreshed into new shares by proactive maintenance, even if viruses are detected and removed from servers' memories. Consequently, the security of (k, n)-threshold protocol against latent virus attacks can be equivalent to that of (k, n)-proactive protocol in case of $\delta = 0$. Therefore, the result of (k, n)-threshold protocol (without proactive maintenance) is an upper bound of a failure probability against this attack.

Let us define $Q(j, t)$ as

$$0 \leq \forall j \leq n \quad Q(j, t) := \Pr[n_i(t) + n_d(t) = j]$$

where j is the number of infected servers. Then, the state transition equations are

$$Q(j, t + \Delta t) = \begin{cases} -n\beta\Delta t \times Q(j, t) & (j = 0) \\ \\ (n - j + 1)\beta\Delta t \times Q(j - 1, t) & \\ \quad -(n - j)\beta\Delta t \times Q(j, t) & (1 \leq j \leq n) \end{cases}$$

Assume at the beginning of the system, the state of all servers to be completely fresh (no infection), namely, $Q(0, 0) = 1$ and $Q(j, 0) = 0$ $(j \geq 1)$. Consequently,

$$Q(j, t) = {}_nC_j \, (e^{\beta t} - 1)^j \, e^{-n\beta t}, \quad P_f(t) = \sum_{j=k}^{n} Q(j, t).$$

A typical threshold is $k = \lfloor n/2 \rfloor$ in most of (k, n)-threshold protocols, because this guarantees the existence of k honest servers and the corruption of no more than $k - 1$ servers (for details, see [1]). Throughout this paper, we assume $k = \lfloor n/2 \rfloor$.

Figure 3 shows the failure probability of $(4, 7)$-threshold protocol. The horizontal axis denotes the progress of time. As one can see clearly, the failure probability of the system increases along with the progress of time. Let us define η as the time when the failure probability $P_f(\eta)$ exceeds $1/2$ at first, which is given as

$$\eta = \ln 2/\beta \approx \beta^{-1}.$$

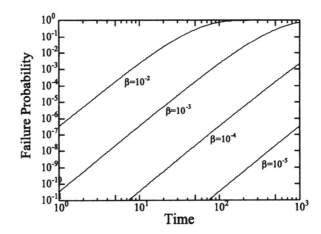

Fig. 3. A failure probability of the $(4, 7)$-threshold protocol

Let us consider after $1000(= 10^3)$ time periods, for example (after about 3 years from the beginning of the system). If $\beta = 10^{-2}$ or $\beta = 10^{-3}$, there is no doubt the system will fail. Even if $\beta = 10^{-4}$, the system is not secure enough to ensure long-term security. It is required that the virus infection rate should not exceed 10^{-5} as long as the system should be available for more than 3 years (in case of the $(4, 7)$-threshold protocol without proactive maintenance).

3.2   Robustness against Latent Virus Attacks (Proactive Protocol)

As figure 3 shows, the assumption of threshold does not hold over a long period of time but within a certain period of time. In threshold protocols without proactive maintenance, once the secret information in servers has leaked out by the viruses, there is no measure to refresh the information and to make the leaked information useless. So, given a sufficient amount of time, the viruses can break into the servers one by one, thus eventually compromising the security of the system. Accordingly, the attacks of the viruses with high infection rate are serious concern in systems that must remain secure for long periods of time.

Proactive maintenance is highly effective against such viruses, due to the mechanism to remove detected viruses and to renew old shares into completely new ones (Figure 1). Therefore, a proactive protocol is more robust against latent virus attacks than a threshold protocol without proactive maintenance. Actually, it is an interesting question how the proactive maintenance enhances the security against the latent viruses with infection rate $\beta = 10^{-2} - 10^{-4}$ which are a threat to $(4, 7)$-threshold protocol without proactive maintenance as mentioned in Section 3.1.

We estimate the above probabilistic model of virus infection and detection in a proactive maintenance. Let us define the probability function $S(j_1, j_2, t)$ as

$$\forall \xi, \forall t \in \mathcal{O}^{(\xi)} \quad S(j_1, j_2, t) := \Pr[n_i(t) = j_2 - j_1, n_d(t) = j_1],$$

where $n - j_2$ servers are uninfected by viruses. $S(j_1, j_2, t)$ means the probabilistic state of the system with respect to virus infection and detection. We can easily see

$$\forall \xi, \forall t \in \mathcal{O}^{(\xi)} \quad P_f(t) = 1 - \sum_{j_2=0}^{k-1} \sum_{j_1=0}^{j_2} S(j_1, j_2, t)$$

At first, let us consider the state transition of the system in an operating phase. There is no change in the direction of decreasing the number of infected servers until the next update phase $\mathcal{U}^{(\xi+1)}$. The system fails when the number of infected servers $n_i(t) + n_d(t)$ exceeds the predetermined threshold $\lceil n/2 \rceil = k - 1$. Accordingly, the state transition of $S(j_1, j_2, t)$ in $\mathcal{O}^{(\xi)}$ is given as, for $\forall \xi$, $\forall t \in \mathcal{O}^{(\xi)}$, $\forall j_1 \forall j_2$ s.t. $0 \le j_1 \le j_2 \le k - 1$,

$S(j_1, j_2, t + \Delta t)$

$$= \begin{cases} -n\beta S(j_1, j_2, t)\Delta t \quad [\text{ if } j_1 = j_2 = 0] \\[2mm] (n - j + 1)\beta S(j_1, j_2 - 1, t)\Delta t - (n - j_2)\beta S(j_1, j_2, t)\Delta t \\ \quad -(j_2 - j_1)\delta S(j_1, j_2, t)\Delta t \quad [\text{ if } j_1 = 0, 1 \le j_2 \le k - 1] \\[2mm] -(n - j_2)\beta S(j_1, j_2, t)\Delta t + \delta S(j_1 - 1, j_2, t)\Delta t \quad [\text{ if } 1 \le j_1 = j_2 \le k - 1] \\[2mm] (n - j_2 + 1)\beta S(j_1, j_2 - 1, t)\Delta t - (n - j_2)\beta S(j_1, j_2, t)\Delta t \\ \quad +(j_2 - j_1 + 1)\delta S(j_1 - 1, j_2, t)\Delta t - (j_2 - j_1)\delta S(j_1, j_2, t)\Delta t \quad [\text{ otherwise }] \end{cases}$$

$$P_f(t + \Delta t) = (n - k + 1)\beta \Delta t \times \sum_{j'=0}^{k-1} S(j', k - 1, t)$$

Secondly, let us consider the state transition of the system in an update phase. The ξ-th update procedure is performed at the beginning of the ξ-th time period, i.e., during $(\xi - 1)\tau \le t \le (\xi - 1)\tau + \tau_u$, where τ_u is the time length of an update phase and it is negligibly short compared with τ. Each transition among $S(j_1, j_2, t)$ is deterministic, that is, all of the state transition probability are equal to 1 or 0. As long as the system does not fail, update procedure renews all shares in the update phases. Consequently, its procedure sets $n_d((\xi-1)\tau+\tau_u)$ at 0 and changes over from $S(j_1, j_2, (\xi-1)\tau)$, i.e., the final state of servers in the $(\xi - 1)$-th operating phase, into $S(0, j_2 - j_1, (\xi-1)\tau +\tau_u)$, i.e., the initial state of servers in the ξ-th operating phase. The state transition equations in the ξ-th update phase $\mathcal{O}^{(\xi)}$ are, for $\forall \xi$, $\forall j_1 \forall j_2$ s.t. $0 \le j_1 \le j_2 \le k - 1$,

$$S(j_1, j_2, \xi\tau + \tau_u) = \begin{cases} \sum_{l=0}^{k-1-j_2} S(l, j_2 + l, \xi\tau) \quad [\text{ if } j_1 = 0] \\[2mm] 0 \quad [\text{ if } j_1 \ne 0] \end{cases}$$

$$P_f(\xi\tau + \tau_u) = P_f(\xi\tau)$$

Now we estimate the security of the system which is engaged in a $(4, 7)$-proactive maintenance after 1000 time periods. Figure 4 shows the relationship between virus infection rate β and detection rate δ achieving each failure probability P_f after 1000 time periods. If the point (δ, β) is located above each of the boundary lines, the system cannot keep the corresponding security against latent virus attacks, otherwise, the system assures the security with less failure probability.

There are remarkable differences on the shape of all lines between $\delta > \delta_t h$ and $\delta < \delta_{th}$, where δ_{th} is the lower bound of effective detection of viruses and in this case (i.e. $(4, 7)$–proactive protocol), we can find $\delta_{th} \simeq 10^{-3}$. From this observation, we can easily see the following requirements for keeping the system secure.

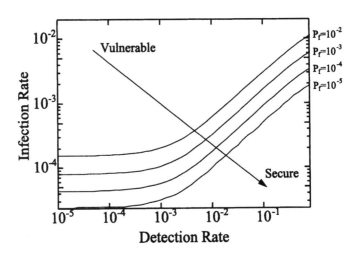

Fig. 4. Virus infection rate against virus detection rate achieving the failure probability $10^{-2}, 10^{-3}, 10^{-4}$ and 10^{-5} after 1000 time periods.

– Case 1 : $\delta > \delta_{th}$
   The requirement for achieving an failure probability P_f is $\beta/\delta < \rho_{th}$, where ρ_{th} is the threshold of infection/detection raito which is determined by the number of servers and the failure probability to be achieved. In this case, the detection and removal of viruses by proactive maintenance works for enhancing security against latent virus attacks.
– Case 2 : $\delta < \delta_{th}$
   The requirement for achieving an failure probability P_f is $\beta < \beta_{th}$, where β_{th} is determined by the number of servers and the failure probability to be achieved. In this case, there is hardly any contribution of proactive maintenance to enhance the security against latent virus attacks because of the lack of the ability to detect the viruses. Therefore, the failure probability of the system practically depends on the infection rate β.

Accordingly, the security of the system against latent virus attacks depends heavily on the ability to detect the infection of the viruses. Especially, in case of $\delta < \delta_{th}$, we can find that the detection and removal of viruses in proactive maintenance does not work effectively enough. Increasing the detection rate can be achieved by improving the virus detection mechanism (anti-virus scanner, etc.), this parameter is too unknown and fluid to estimate the security of the system appropriately. Therefore, from the viewpoint of constructing mission critical system, it is to be desired that we could cope with more powerful viruses by more controllable means even in a case where the ability to detect viruses is not enough. The method to meet these demands is "active rebooting method" which we present in the following session.

4   Active Rebooting

Above proactive maintenance does not work enough in case of low detection rate. This is caused by the fact that the server is not rebooted until it becomes aware of the existence of viruses (passive rebooting). Therefore, the unawareness of viruses intruding and being latent in servers results in neglecting to take appropriate measures promptly. Accordingly, we propose a new method of rebooting servers, which we call "active rebooting" method . Active rebooting assures us that in each update phase, more than a predetermined number α of servers (of course, $\alpha < k$) is rebooted and their states were made uninfected. Since active rebooting method does not depend on detection of viruses for their removal, it is expected to remarkably reduce the failure probability especially in case of low detection rate.

We denote the probability that in case of $n_i(\xi\tau) = j_2 - j_1$, $n_d(\xi\tau) = j_1$, j_0 servers are additionally rebooted by active rebooting in case of as $R(j_0, j_1, j_2)$, which can be formally defined as, for $\forall\xi$, $\forall j_0$, $\forall j_1 \forall j_2$ s.t. $0 \leq j_1 \leq j_2 \leq k - 1$,

$$R(j_0, j_1, j_2) = \Pr[\, n_i(\xi\tau) = j_2 - j_1, \; n_d(\xi\tau) = j_1, \; n_i(\xi\tau + \tau_u) = j_1 - j_0 \,].$$

If $\alpha \leq j_1$ $(< k)$, no additional servers are rebooted except servers in $\mathcal{D}(\xi\tau)$ since $n_d(\xi\tau)$ exceeds α. So, undetected viruses are still in the infected servers. Accordingly, if $\alpha \leq j_1$ and $j_0 = 0$, $R(j_0, j_1, j_2) = 1$, while, if $\alpha \leq j_1$ and $j_0 \neq 0$, $R(j_0, j_1, j_2) = 0$.

If $0 \leq j_1 < \alpha$, $\alpha - j_1$ out of $n - j_1$ servers in $\bar{\mathcal{D}}(\xi\tau)$ (referred to as $\mathcal{R}$) are additionally rebooted and the total number of rebooted servers are α. In this case, $j_2 - j_1$ out of $n - j_1$ servers are infected and the other $n - j_2$ servers are uninfected. $R(\cdot)$ is equivalent to the probability that j_0 infected servers out of $n - j_1$ servers in $\bar{\mathcal{D}}(\xi\tau)$ are included in $\mathcal{R}$. Accordingly,

$$R(j_0, j_1, j_2) = \frac{{}_{j_2-j_1}C_{j_0} \times {}_{n-j_2}C_{\alpha-j_1-j_0}}{{}_{n-j_1}C_{\alpha-j_1}} \; [\text{ if } 0 \leq j_1 < \alpha, \; 0 \leq j_1 \leq j_2 < n/2,$$
$$0 \leq j_0 \leq \min(\alpha - j_1, j_2 - j_1) \,],$$

otherwise, $R(j_0, j_1, j_2) = 0$. Using the probabilistic function $R(\cdot)$, for $\forall\xi$, $\forall j_1 \forall j_2$ s.t. $0 \leq j_1 \leq j_2 \leq k - 1$, the state transition shown in section 3.2 is replaced as

$$S(j_1, j_2, \xi\tau+\tau_u) \begin{cases} = \sum_{l_1=0}^{k-1-j_2} \sum_{l_2=0}^{k-1-j_2-l_1} S(l_2, j_2 + l_2 + l_1, \xi\tau) \\ \qquad\qquad \times R(l_1, l_2, j_2 + l_1 + l_2) \quad [\text{ if } j_1 = 0] \\[2ex] = 0 \quad [\text{ if } j_1 \neq 0] \end{cases}$$

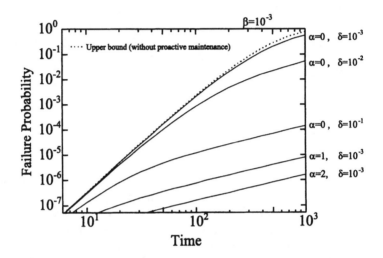

Fig. 5. A failure probability with respect to the progress of time for $\beta = 10^{-3}$.

These are the state transition formulas of the system which is engaged in a (k, n)-proactive protocol with active rebooting in the ξ-th update phase. In the same way as the previous section, we estimate the security of the system which is engaged in a $(4, 7)$-proactive maintenance with active rebooting. Figure 5 shows the failure probability $P_f(t)$ with respect to the progress of time for $\beta = 10^{-3}$. The vertical axis means the failure probability and the horizontal axis means the progress of time. A dotted line is the upper bound of failure probability of the $(4, 7)$-threshold protocol (See section 3.1.) If the detection rate $\delta < 10^{-2}$, there is no improvement even though the system performs the proactive maintenance. Applying active rebooting to update procedure enables us to reduce the failure probability even in case of low detection ability. For example, the system to which the active rebooting is applied in case of $\alpha = 1, \delta = 10^{-3}$ has a robustness against latent virus attacks than passive rebooting system in case of higher detection rate $\delta = 10^{-1}$. We remark that active rebooting is effective as compared with passive rebooting even when small α, that is, the increase of processing is comparatively small.

Figure 6 shows the relationship between the infection rate β and the detection rate δ achieving the failure probability $P_f = 10^{-3}$ after 1000 time periods. If the point (δ, β) is located above each of the boundary lines, the system cannot keep the corresponding security against latent virus attacks, otherwise, the system

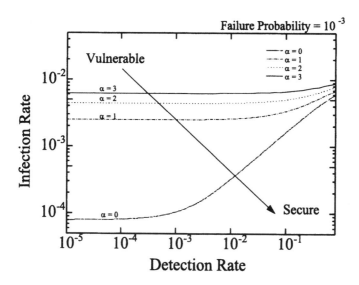

Fig. 6. Virus infection rate against virus detection rate achieving the failure probability 10^{-3} after 1000 time periods.

assures the security with less failure probability. The curve of $\alpha = 0$ (without active rebooting) is the same as figure 4.

Employing the active rebooting enhances robustness against latent virus attack with higher infection rate β throughout the entire range of δ, so that the failure probability is not so dependent on the detection rate δ. We mention that the proactivized system with active rebooting even in the case of relatively small δ can assure the same level of security against the latent viruses as that of high δ. This is caused by the mechanism of active rebooting removing infected and hidden viruses in servers. This result means that the use of active rebooting is highly effective against the attack of latent viruses, even though β and δ may not be measured quantitatively in practice. Therefore, it is important to use proactivized system in an active rebooting manner in order to maintain the security against unknown viruses.

5   Conclusions

In this paper, we proposed the method for enhancing the robustness of proactive protocol against latent virus attacks. Our contribution in this paper is as follows.

At first, we examined the robustness of proactive system against the attack caused by latent viruses. We presented some tradeoff among an infection rate of viruses, a detection rate of servers and a failure probability of the system in which a certain number of servers engages in a proactive protocol. As a result, we showed the fact that if the virus detection rate is higher than a certain threshold, it is possible for the proactive maintenance to make the system robust, while, if less than the threshold, the failure probability of the system is dependent only on the virus infection rate.

Secondly, we proposed the notion of *active rebooting*, in which the system performs the reboot procedure on a predetermined number of servers in the total independence of servers being infected or not. We estimated the security of proactive maintenance with active rebooting by extending the probabilistic model of proactive maintenance. We showed that active rebooting enables us not only to enhance the security against the viruses with higher infection rate, but also to make the system robust even in the case of a low detection rate. Moreover, we showed that it is effective even in the case the number of servers which are forced to carry out the reboot operation every update phase is comparatively small.

Arranging the procedure of active rebooting in a FIFO (First In First Out) manner, we can set reasonable limitation on the period in which virus is able to be latent in a server. We can use this fact as new attacker's model in order to enhance the security against virus attacks. Timed-release cryptosystem[16][17] is one of the candidates for application. Forthcoming work will show how to apply these results to timed-release cryptosystem.

Acknowledgements

We thank Kazukuni Kobara and Motohiko Isaka for interesting and helpful comments. Special thanks to Hideki Ochiai and Joern Mueller Quade for valuable comments and important guidance during the preparation of the final version. Finnaly, we wish to thank the anonymous referees for important feedback.

References

1. R. Ostrovsky and M. Yung. How to withstand mobile virus attacks. In *Proc. of PODC'91*, pages 51–59, 1991.
2. A. Herzberg, S. Jarecki, H. Krawczyk, and M. Yung. Proactive secret sharing, or: How to cope with perpetual leakage. In *Proc. of CRYPTO'95*, pages 339–352, 1995.
3. A. Herzberg, M. Jakobsson, S. Jarecki, and H. Krawczyk. Proactive public key and signature systems. In *Proc. of The 4-th ACM Symposium on Computer and Communication Security'97*, April 1997.
4. Y. Frankel, P. Gemmell, P. Mackenzie, and M. Yung. Proactive RSA. In *Proc. of CRYPTO'97*, pages 440–454, 1997.
5. A. Shamir. How to share a secret. *Comm. of ACM*, 22:612–613, 1979.
6. Y. Desmedt. Threshold cryptosystem. *European Transactions on Telecommunications*, 5(4):449–457, 1994.
7. P. S. Gemmell. An introduction to threshold cryptography. *CryptoBytes*, 2(3):7–12, 1997.
8. Y. Frankel and M. Yung. Distributed public key cryptography. In *Proc. of PKC'98*, pages 1–13, 1998.

9. R. Canetti, R. Gennaro, A. Herzberg, and D. Naor. Proactive security: Long-term protection against break-ins. *CryptoBytes*, 3(1):1–8, 1997.
10. P. Feldman. A practical scheme for non-interactive verifiable secret sharing. In *Proc. of FOCS'87*, pages 427–437, 1987.
11. T. P. Pedersen. Non-interactive and information-theoretic secure verifiable secret sharing. In *Proc. of CRYPTO'91*, pages 129–140, 1991.
12. F. Cohen. Computer viruses, theory and experiments. *Computers & Security*, 6:22–35, 1987.
13. Y. Sengoku, E. Okamoto, M. Mambo, and T. Uematsu. Analysys of infection and distinction of computer viruses in computer networks. In *Proc. of International Symposium on Information Theory and Its Applications (ISITA'96)*, pages 163–166, 1996.
14. J. Kephart and S. White. Directed-graph epidemiological models of computer viruses. In *Proc. of IEEE Symposium on Security and Privacy*, pages 343–359, 1991.
15. A. De Santis, Y. Desmedt, Y. Frankel, and M. Yung. How to share a function securely. In *Proc. of STOC'94*, pages 522–533, 1994.
16. R. L. Rivest, A. Shamir, and D. A. Wagner. Time-lock puzzles and timed-release crypto. In *Manuscript at* http://theory.lcs.mit.edu/~ rivest/.
17. G. D. Crescenzo, R. Ostrovsky, and S. Rajagopalan. Conditional oblivious transfer and timed-release encryption. In *Proc. of Eurocrypt'99*, pages 74–89, 1999.
18. L. Adleman. Abstract theory of computer viruses. In *Proc. of CRYPTO'88*, pages 354–374, 1988.

Highly Robust Image Watermarking Using Complementary Modulations

Chun-Shien Lu, Hong-Yuan Mark Liao, Shih-Kun Huang, and Chwen-Jye Sze

Institute of Information Science,
Academia Sinica, Taipei, Taiwan
{lcs, liao}@iis.sinica.edu.tw

Abstract. In this paper, a novel image protection scheme using "complementary modulations" is proposed. In order to satisfy the robustness requirement, two watermarks are embedded into a host image to play complementary roles so that at least one watermark survives any attack. Statistical analysis is conducted to derive the lower bound of the worst likelihood that the better watermark (of the two) can be extracted. With this "high" lower bound, it is ensured that a "better" extracted watermark is always obtained. Experimental results demonstrate that our watermarking scheme is remarkably effective in resisting various attacks.

1   Introduction

Use of the Internet for transferring digitized media has become very popular in recent years. However, this frequent use of the Internet has created a need for security. Therefore, to prevent information which belongs to rightful owners from being intentionally or unwittingly used by others, information protection is indispensable. A commonly used method is to insert watermarks into original information so that rightful ownership can be declared. This is the so-called watermarking technique. An effective watermarking procedure usually requires satisfaction of a set of typical requirements. These requirements include transparency (perceptual invisibility), robustness, maximum capacity, universality, oblivious watermarking, and resolution of ownership deadlock [4].

It is well known that the current watermarking approaches are not robust to attacks, so their use is limited [8]. In this paper, the above mentioned problem will be seriously addressed. To better understand the nature of the problem, we shall begin by introducing two commonly referred works. The first one is the spread spectrum watermarking technique proposed by Cox *et al.* [3]. The method has become very popular and has been employed by many researchers [2,6,7,15]. The other one, proposed by Podilchuk and Zeng [14], is a human visual model-based watermarking scheme. Their work has also been extensively cited [5,6,14,16]. However, the reasons why the two aforementioned methods are successful or not are still not clear. Here, the modulation techniques used in [3,14] will be investigated. We assert that in order to obtain high detector responses, most of the transformed coefficients of the host image and the watermarked image have to be modulated along the same direction. This is the

M. Mambo, Y. Zheng (Eds.): ISW'99, LNCS 1729, pp. 136–153, 1999.
© Springer-Verlag Berlin Heidelberg 1999

key concept needed to improve the previous approaches because a watermark detector can produce a high correlation value only when the above mentioned condition is satisfied. Unfortunately, we find that both Cox *et al.*'s method [3] and Podilchuk and Zeng's method [14] do not take this important factor into account. We have observed that an arbitrary attack either decreases or increases the magnitudes of the majority ($> 50\%$) of the transformed coefficients. In other words, the chance that an attack will make the number of increased and of decreased coefficients equal is very rare. In this paper, we propose an efficient modulation strategy, which is composed of a positive modulation (increasing the magnitude of transformed coefficients) and a negative modulation (decreasing the magnitude of transformed coefficients). The two modulation rules simultaneously hide two complementary watermarks in a host image so that at least one watermark survives under different attacks. We have also conducted statistical analysis to derive a lower bound, which provides the worst likelihood that the better watermark (of the two) can be extracted. With this bound, a "better" extracted watermark is always obtained. Experimental results confirm that our watermarking scheme is extremely robust to different kinds of attacks.

2   Watermark Modulation

In the transformed domain, watermark modulation is an operation that alters the values of selected transformed coefficients using every selected coefficient's corresponding watermark value. We will analyze and point out the inadequacy of the modulation techniques commonly used in ordinary spread spectrum watermarking methods and the visual model-based ones. To resolve this inadequacy, two watermarks which play complementary roles are simultaneously embedded into a host image by means of the proposed "complementary modulations."

In what follows, we shall first introduce the basic concept of random modulation. The proposed complementary modulations will be described in detail in Sec. 2.2.

2.1   Random Modulation

Two very popular watermarking techniques, which take the spread spectrum concept and the perceptual model into account, were presented in [3,14]. Cox *et al.* [3] used the spread spectrum concept to hide a watermark based on the following modulation rule:

$$I_i^* = I_i(1 + \alpha \cdot n_i),   (1)$$

where I_i and I_i^* are DCT coefficients before and after modulation, respectively, and n_i is a value of a watermark sequence. α is a weight that controls the trade-off between transparency and robustness. In [14], Podilchuk and Zeng presented two watermarking schemes based on the human visual model, i.e., the image adaptive-DCT (IA-DCT) and the image adaptive wavelet (IA-W) schemes. The watermark encoder for both IA-DCT and IA-W can be generally described as

$$I_{u,v}^* = \begin{cases} I_{u,v} + J_{u,v} \cdot n_{u,v}, & I_{u,v} > J_{u,v}; \\ I_{u,v}, & otherwise, \end{cases} \tag{2}$$

where $I_{u,v}$ and $I_{u,v}^*$ are DCT or wavelet coefficients before and after modulation, respectively, and $n_{u,v}$ is the sequence of watermark values. $J_{u,v}$ is the masking value of a DCT or a wavelet based visual model. Basically, this masking value is the so-called Just Noticeable Distortion (JND) mentioned in [17]. It is found from both embedding schemes that modulations take place in the perceptually significant coefficients with the modification quantity specified by a weight. The weight is either heuristic [3] or depends on a visual model [14]. Cox *et al.* [3] and Podilchuk and Zeng [14] both adopted a similar detector response measurement described by

$$Sim(n, n^*) = \frac{n \cdot n^*}{\sqrt{n^* \cdot n^*}}, \tag{3}$$

where n and n^* are the original and the extracted watermark sequences, respectively. If the signs of a corresponding pair of elements in n and n^* are the same, then they contribute positively to the detector response. A higher value of $Sim(n, n^*)$ means stronger evidence that n^* is a genuine watermark. In Eq. (3), high correlation values can only be achieved if most of the transformed coefficients of the original source and the watermarked image are updated along the same direction during the embedding and the attacking processes, respectively. This is the key point if a watermark detector is to get a higher correlation value. However, we find that neither [3] nor [14] took this important factor into account. In fact, the modulation strategy they adopted is intrinsically random. Usually, a positive coefficient can be updated with a positive or negative quantity, and a negative coefficient can be altered with a positive or a negative quantity as well. In other words, the [3] and [14] did not consider the relationship between the signs of a *modulation pair*, which is composed of a selected transformed coefficient and its corresponding watermark value. This explains why many attacks can successfully defeat the above mentioned watermarking schemes.

In the following analysis, we assume that the watermark sequence n is embedded into a host image H. For the random modulation techniques proposed in [3] and [14], there are four possible types of modulations: $Modu(+, +)$, $Modu(+, -)$, $Modu(-, +)$, and $Modu(-, -)$, where $Modu(+/-, -/+)$ represents a positive/negative transformed coefficient modulated with a negative/positive watermark quantity. For a noise-style watermark with a Gaussian distribution of zero mean and unit variance, the probability of drawing a positive or a negative value is roughly equal to 0.5. In the wavelet domain, the wavelet coefficients of a high-frequency band can be modeled as a generalized Gaussian distribution [1] with the mean close to 0; i.e., the probability of getting a positive or a negative coefficient is roughly equal to 0.5. The lowest frequency component is, however, only suitably modeled by a usual Gaussian distribution with the mean far away from 0. That is, the probability of obtaining a positive coefficient is extremely different from that of obtaining a negative coefficient. When wavelet decomposition is executed with many scales, the lowest frequency component is tiny. Therefore, the probability of getting a positive or a negative wavelet coefficient is still close to 0.5. For the transformed coefficients in the DCT domain, the numbers of the positive and negative global DCT coefficients are statistically very close to each other. Hence, no matter whether DCT or the wavelet domain is employed, the probabilities of occurrence of the four types of modulations are all very close to 0.25 due to their characteristic of randomness. We have also analyzed the influence of a number of attacks to see how they update the magnitude of each transformed coefficient. From our observations, we find that by using the random modulation proposed in [3,14], about 50% of the transformed coefficients can be increasingly modulated and the other half are decreasingly modulated. Therefore, it can be concluded that the random modulation strategy does not help to increase the detector response value at all.

For an attack like sharpening, the magnitudes of most of the transformed coefficients of a watermarked image increase when it is applied. Under these circumstances, it is hoped that every transformed coefficient can be modulated with a quantity that has the same sign. The reason why the above modulation strategy is adopted is that it can adapt to sharpening-style attacks and enable more than 50% of the modulated targets to contribute a bigger positive value to the detector response. In sum, we can conclude that of the four types of modulations, only $Modu(+,+)$ and $Modu(-,-)$ will contribute a bigger positive value to the detector response. On the other hand, if an attack (ex., blurring) causes most of the transformed coefficients to decrease in magnitude, then every constituent transformed coefficient should be modulated with a quantity that has a different sign. Under these circumstances, only $Modu(+,-)$ and $Modu(-,+)$ will contribute a bigger positive value to the detector response.

2.2 Complementary Modulations

From the analysis described in Sec. 2.1, it is clear that a good watermarking scheme should be adaptive to various types of attacks. In this section, we shall propose a new modulation scheme which can resist different kinds of attacks. It is noted that the detector response defined in Eq. (3) is a function of n and n^*. Basically, n is the sequence of watermark values and is, therefore, fixed once it is chosen. However, the values of n^* are dependent on the strength of an attack. Since we are concerned with the direction of modulation instead of the amount of change [12], the watermark value is defined in the bipolar form, that is,

$$bipolar(t) = \begin{cases} 1, & t \geq 0 \\ -1, & t < 0, \end{cases} \tag{4}$$

where t is a real number. The existence of a watermark pixel is, thus, determined by the sign of a piece of retrieved information. Under these circumstances, even if a watermarked image is hit by a very strong attack, the resultant detector response will not drop dramatically. Since the watermark value adopted is bipolar, the detector response for bipolar watermarks proposed by Kundur and Hatzinakos [9] is used:

$$\rho(n, n^e) = \frac{\sum n(i) n^e(i)}{L_M},$$ (5)

where $n(i)$ ($i = 1, 2, ..., L_M$) is the sequence of embedded watermark values, $n^e(i)$ is the extracted watermark values, and L_M is the length of the hidden watermark. The extracted watermark values, n^e, can be determined by means of the bipolar test described in Eq. (4). If a watermark image has been attacked and the coordinates in the transformed domain are (x, y), then the extracted watermark can be expressed as

$$
\begin{aligned}
n^e(i) = n^e(m(x, y)) &= bipolar(H^a(x, y) - H(x, y)) \\
&= bipolar((H^a(x, y) - H^m(x, y)) + (H^m(x, y) - H(x, y))) \\
&= bipolar(\beta_1 + \beta_2), \quad i = 1, 2, ..., L_M,
\end{aligned}
$$ (6)

where $H(x, y)$, $H^m(x, y)$, and $H^a(x, y)$ represent the original, the modulated, and the attacked transformed coefficients, respectively, detected at (x, y) in the transformed domain. The mapping function m forms a one-to-one mapping (will be established in Sec. 3) which maps a selected transformed coefficient to its corresponding watermark index. From the analysis described in Sec. 2.1, it is clear that in order to obtain a high detector response, the signs of $n(i)$ and $n^e(i)$ have to be the same. There exists two possible conditions under which $n(i)$ and $n^e(i)$ will have the same sign. First, if β_1 and β_2 have the same sign, then $bipolar(\beta_1 + \beta_2)(= n^e(i))$ and $bipolar(\beta_2)(= n(i))$ will be the same. A complementary modulation strategy is proposed to satisfy this requirement. The scenario 1 of Fig. 1 illustrates this phenomenon. The second condition is that β_1 and β_2 have different signs, but that $|\beta_1| < |\beta_2|$. Under these circumstances, the modulated amount is larger than the amount altered by an attack. In other words, the applied attack is not strong enough to influence the sign change created by the modulation process. This situation is illustrated in the scenario 2 of Fig. 1. Introduction of the second condition is necessary to obtain a higher detector response because it intrinsically makes use of the masking effect of the human visual model and, thus, maximizes the hiding capacity. In this paper, the wavelet-based visual model [17] is adopted to help us determine the maximum capacity that is allowed to embed watermarks.

In what follows, a complementary modulation strategy is presented which can adapt to different kinds of attacks. The proposed scheme embeds two watermarks, which play complementary roles in resisting various kinds of attacks. The two watermarks are embedded using two different modulation rules: positive modulation and negative modulation. If a modulation operates by adding a negative quantity to a positive coefficient or by adding a positive quantity to a negative coefficient, then we call it "negative modulation." Otherwise, it is called "positive modulation" if the sign of the added quantity is the same as that of the corresponding transformed coefficient. The robustness demand is always guaranteed since at least one of the two watermarks is able to capture the behavior of an encountered attack.

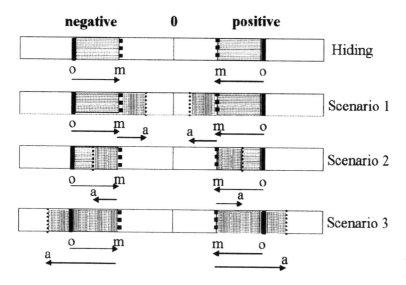

Fig. 1. Scenarios in the attacking process for negative modulation. 'o' denotes the original wavelet coefficient, 'm' represents the wavelet coefficient after modulation, and 'a' is the coefficient after attacks; positive/negative denotes the part of positive/negative wavelet coefficients; the areas with horizontal/vertical dashed lines represents the magnitude updated by the hiding/attacking process; the arrow from 'o' to 'm'/'m' to 'a' indicates the direction of the modulation/attacking process: (top) hiding using negative modulation; (scenario 1) the behaviors of the hiding and the attacking processes are the same; (scenario 2/scenario 3) the behaviors of the hiding and the attacking processes are different but the strength of the attack is weaker/stronger than that of the negative modulation.

Let R_{nm}^M be a set of locations in the transformed domain whose corresponding transformed coefficients are to be decreased in magnitude, and let $H_{s,o}(x_h, y_h)$ and $H_{s,o}^m(x_h, y_h)$ be the original and the modulated wavelet coefficients, respectively, at (x_h, y_h). The subscripts s and o represent, respectively, scale and orientation. The explicit form of R_{nm}^M can be expressed as follows:

$$
\begin{aligned}
R_{nm}^M &= \{(x_h, y_h) | where \ (H_{s,o}^m(x_h, y_h) - H_{s,o}(x_h, y_h)) \cdot H_{s,o}(x_h, y_h) < 0\} \\
&= \{(x_h, y_h) | |H_{s,o}^m(x_h, y_h)| < |H_{s,o}(x_h, y_h)|\}.
\end{aligned} \tag{7}
$$

The embedding rule that specifies the condition $(H_{s,o}^m(x_h, y_h) - H_{s,o}(x_h, y_h)) \cdot H_{s,o}(x_h, y_h) < 0$ is called "**negative modulation (NM)**." The set R_{nm}^M is altered and becomes a new set, R_{nm}^{M*}, after an attack. The set of elements R_{nm}^A, which indicates the locations where the embedding and the attacking processes behave consistently should be identified. This set can be expressed as follows:

$$R_{nm}^A = R_{nm}^M \cap R_{nm}^{M*}$$
$$= \{(x_h, y_h) | where \ (H_{s,o}^a(x_h, y_h) - H_{s,o}^m(x_h, y_h)) \cdot H_{s,o}^m(x_h, y_h) < 0\}$$
$$= \{(x_h, y_h) | n(m(x_h, y_h)) \cdot n^e(m(x_h, y_h)) > 0\}, \tag{8}$$

where $H_{nm}^a(x_h, y_h)$ is the attacked wavelet coefficient. Since the modulation and attack processes behave in the same way at (x_h, y_h), $n(m(x_h, y_h)) \cdot n^e(m(x_h, y_h)) > 0$ holds and contributes to the detector response. On the other hand, a "**positive modulation (PM)**" event for watermark encoding can be defined as $n(m(x_h, y_h)) \cdot H_{s,o}(x_h, y_h) > 0$. Therefore, the set of locations whose corresponding coefficients are increasingly modulated in magnitude, P_{pm}^M, can be defined as

$$R_{pm}^M = \{(x_h, y_h) | (H_{s,o}^m(x_h, y_h) - H_{s,o}(x_h, y_h)) \cdot H_{s,o}(x_h, y_h) > 0\}$$
$$= \{(x_h, y_h) | |H_{s,o}^m(x_h, y_h)| > |H_{s,o}(x_h, y_h)|\}. \tag{9}$$

The set R_{pm}^A, which contains locations where the wavelet coefficients are increasingly modulated in magnitude by an attack given that a positive modulation event has occurred, can be represented as

$$R_{pm}^A = \{(x_h, y_h) | (H_{s,o}^a(x_h, y_h) - H_{s,o}^m(x_h, y_h)) \cdot H_{s,o}^m(x_h, y_h) > 0\}$$
$$= \{(x_h, y_h) | n(m(x_h, y_h)) \cdot n^e(m(x_h, y_h)) > 0\}. \tag{10}$$

Notice that only one watermark is hidden with respect to each modulation rule (event) under this complementary modulation strategy. It is obvious that the two sets R_{nm}^M and R_{pm}^M are disjointed. That is,

$$R_{nm}^M \cap R_{pm}^M = \emptyset.$$

For an attack that favors negative modulation, the magnitude of most ($>$ 50%) of the transformed coefficients will be decreased. Let P_{nm}^A be the probability that wavelet coefficients will be decreasingly modulated (in magnitude) by an attack provided that the embedding rule "**negative modulation**" has occurred. It is defined as

$$P_{nm}^A = \frac{|R_{nm}^A|}{|R_{nm}^M|}, \tag{11}$$

where $|S|$ denotes the number of elements in the set S. Ideally, the condition $P_{nm}^A = 1$ only holds for an attack whose behavior completely matches negative modulation. That is, all the coefficients of the original image and the watermarked image are decreased. In fact, it is difficult for an attack to match the behavior of negative modulation completely. Therefore, the relation $|R_{nm}^A| \leq |R_{nm}^M|$ holds. Furthermore, under the assumption that the attack favors negative modulation, $\frac{1}{2}|R_{nm}^M| < |R_{nm}^A|$ holds. That is,

$$\frac{1}{2}|R_{nm}^M| < |R_{nm}^A| \leq |R_{nm}^M|, \tag{12}$$

and

$$P_{nm}^A \in (0.5 \ 1]. \tag{13}$$

From Eq. (13), we know that more than 50% of the pairs of $(n(\cdot,\cdot), n^e(\cdot,\cdot))$ will have the same sign and, thus, contribute to the detector response. These pairs result from the fact that more than 50% of the transformed coefficients are decreased in their magnitudes. Similar procedures can be deduced to obtain P^A_{pm} given that positive modulation has occurred. One may ask what will happen if we do not know the tendency of an attack in advance. Fortunately, since our approach hides two complementary watermarks in a host image, at least one modulation will match the behavior of an arbitrary attack with the probability, P^A, guaranteed to be larger than 0.5; i.e.,

$$P^A = MAX\{P^A_{nm}, P^A_{pm}\} > 0.5. \tag{14}$$

3   The Proposed Watermarking Algorithm

Currently, our host images are gray-scale. The wavelet transform adopted here is constrained such that the size of the lowest band is 16×16. Our bipolar watermark is randomly generated as the signs of a Gaussian distribution of zero mean and unit variance and the magnitudes of the Gaussian sequence are used as the weights for modulation.

3.1   Wavelet Coefficients Selection

The area for hiding watermarks is divided into two parts, i.e., a lowest frequency part and a part that covers the remaining frequencies. It is noted that the lowest frequency wavelet coefficients correspond to the largest portion of a decomposition. In this approach, different weights may be assigned to achieve a compromise between transparency and robustness. Here, only the frequency masking effect of the wavelet-based visual model [14,17] is considered. Owing to the lack of wavelet-based image-dependent masking effects, the heuristic weights assignment needs to be used.

Before the wavelet coefficients of the host images are modulated, we must select the places for embedding. The wavelet coefficients are selected for modulation if their magnitudes are larger than their corresponding JND thresholds. Because two complementary watermarks need to be hidden, the length of each watermark to be hidden is one half the length of the selected coefficients. Basically, a watermark designed using our approach is image-adaptive [14]. Next, the places in the wavelet domain should be selected for watermarks hiding in an interleaving manner. We use a secret key to generate a Gaussian sequence, G, with zero mean and its length equal to the number of selected wavelet coefficients. The selected wavelet coefficients and this Gaussian form a one-to-one mapping. This mapping function is defined as

$$m(x,y) = \begin{cases} 1, & G(i) \geq 0 \\ -1, & G(i) < 0, \end{cases} \tag{15}$$

where (x, y) is the coordinate in the wavelet domain and i is the index of this Gaussian sequence, G. The locations in the wavelet domain correspond to positive/negative values will be assigned to employ positive/negative modulation rules. Our complementary modulations described in the next section are, then, conducted after the wavelet coefficients selection.

3.2   Watermark Hiding

The watermark embedding process proceeds as follows. The wavelet coefficients (H) of a host image are sorted in increasing order according to their absolute values and the watermark sequence $(\mathcal{N})$ is sorted in increasing order, too. Each time, a pair of wavelet coefficients $H_{s,o}(x_p, y_p)$ and $H_{s,o}(x_n, y_n)$ is selected from the top of the sorted coefficient sequence. They are then modulated and become $H_{s,o}^m(x_p, y_p)$ and $H_{s,o}^m(x_n, y_n)$, respectively, according to the following modulation rules:

Positive modulation:

$$H_{s,o}^m(x_p, y_p) = \begin{cases} H_{s,o}(x_p, y_p) + J_{s,o}(x_p, y_p) \cdot bipolar(n_{bottom}) \cdot |n_{bottom} \cdot w|, \\ \qquad\qquad\qquad\qquad\qquad H_{s,o}(x_p, y_p) \geq 0 \\ H_{s,o}(x_p, y_p) + J_{s,o}(x_p, y_p) \cdot bipolar(n_{top}) \cdot |n_{top} \cdot w|, \\ \qquad\qquad\qquad\qquad\qquad H_{s,o}(x_p, y_p) < 0, \end{cases} \quad (16)$$

where $J_{s,o}(.,.)$ represents the JND values [17] of a wavelet-based visual model and n_{top}/n_{bottom} represents the value retrieved from the top/bottom of the sorted watermark sequence $\mathcal{N}$. $bipolar(\cdot)$ serves as the bipolar watermark value. w is a weight used to control the maximum possible modification that will lead to the least image quality degradation. It is defined as

$$w = \begin{cases} w_L, & H_{s,o}(\cdot, \cdot) \in lowest - frequency\ channel \\ w_H, & H_{s,o}(\cdot, \cdot) \in high - frequency\ channel. \end{cases} \quad (17)$$

w_L and w_H refer to the weights imposed on the low and high frequency coefficients, respectively. If both of them are set to be one, they are diminished as in [14].

Negative modulation:

$$H_{s,o}^m(x_n, y_n) = \begin{cases} H_{s,o}(x_n, y_n) + J_{s,o}(x_n, y_n) \cdot bipolar(n_{top}) \cdot |n_{top} \cdot w|, \\ \qquad\qquad\qquad\qquad\qquad if\ H_{s,o}(x_n, y_n) \geq 0 \\ H_{s,o}(x_n, y_n) + J_{s,o}(x_n, y_n) \cdot bipolar(n_{bottom}) \cdot |n_{bottom} \cdot w|. \\ \qquad\qquad\qquad\qquad\qquad if\ H_{s,o}(x_n, y_n) < 0 \end{cases} \quad (18)$$

Based on the above mentioned positive and negative modulations, the mapping relationship between the position of a selected wavelet coefficient and the index of its corresponding watermark value is established as

$$m(x, y) = \begin{cases} i, & G(i) \geq 0 \\ -i, & G(i) < 0. \end{cases} \quad (19)$$

These mapping results will be stored for watermark detection and kept secret such that the pirates cannot easily remove the hidden watermarks. So, in the watermark detection process, we search for the positive/negative *signs* of $m(x, y)$ to detect watermarks embedded by positive/negative modulation rules. Besides, the positive/negative *values* of $m(x, y)$ determine the index of hidden watermarks. Fig. 2 illustrates the whole process of our watermark hiding technique.

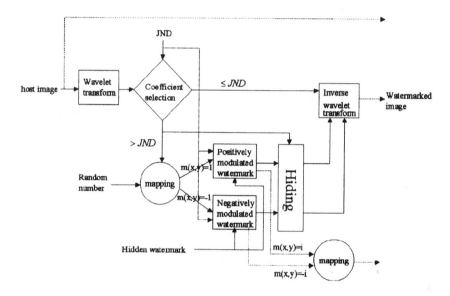

Fig. 2. The watermark encoding process of our cocktail watermarking scheme.

3.3 Watermark Detection

In the literature, a number of authors [2,6,7,9,10] have proposed use of the public mode to extract a watermark without access to the original image, but the correlation values detected using their methods are not high enough, especially under strong attacks. For instance, Kutter *et al.* [10] predicted an original DCT coefficient based on the distorted DCT coefficients in a local region. Barni *et al.* [2] skipped the largest N DCT coefficients and tried to decorrelate the low-frequency part of a host image and the extracted watermark. To eliminate cross-talk between the video signal and the watermark signal, Hartung *et al.* [7] applied high-pass filtering to the attacked watermarked video. On the other hand, the information about a distorted image can be used directly as if it comes from the original image [6,9]. It is obvious that the performance of these public modes is not good enough due to the lack of a precise way to predict the original image. Currently, the original image is still needed to extract watermarks due to the lack of a reliable oblivious watermarking technique. Basically, the need of host image is suitable for destination-based watermarking [14].

Dealing with Attacks Which Include Asynchronous Phenomena In this section, we shall present a relocation strategy we can use to tackle attacks that generate asynchronous phenomena. In what follows, we shall introduce some attacks of this sort. StirMark [13] is a very strong type of attack that defeats many watermarking techniques. Analysis of StirMark [13] has shown that it introduces unnoticeable quality loss in an image with some simple geometrical distortions. Jitter [13] is another type of attack, which leads to spatial errors in images that are perceptually invisible. Basically, these types of attack cause asynchronous problems. Experience tells us that an embedded watermark, which encounters these attacks, is often severely degraded [11]. Therefore, it is important to deal with the attack in a clever way so that damage can be minimized. It is noted that the order of wavelet coefficients is different before and after an attack and might vary significantly under attacks with the inherent asynchronous property. Consequently, in order to recover a "correct" watermark, the wavelet coefficients of an attacked watermarked image should be relocated to their proper positions before watermark detection is conducted. The relocation operation is described in the following. First, the wavelet coefficients of the attacked watermarked image are re-arranged into the same order as those of the watermarked image. Generally speaking, by preserving the orders, damage to the extracted watermark can always be reduced. In the experiments, we shall see that the detection response performance measured in the relocation step is significantly improved. The improvement in the results is especially remarkable for some attacks.

Determination of Detector Response According to the mapping function, the detector responses resulting from positive modulation and negative modulation are represented by $\rho^{pos}(\cdot, \cdot)$ and $\rho^{neg}(\cdot, \cdot)$, respectively. The final detector response, $\rho^{CM}(\cdot, \cdot)$, is thus defined as

$$\rho^{CM}(\cdot, \cdot) = MAX(\rho^{pos}(\cdot, \cdot), \rho^{neg}(\cdot, \cdot)), \qquad (20)$$

where CM is an abbreviation of Complementary Modulations. Furthermore, if the relocation step is applied, then the detector response is denoted as $\rho^{CM}_{Re}(\cdot, \cdot)$; otherwise, it is denoted as $\rho^{CM}_{NRe}(\cdot, \cdot)$. A better detector response can be determined by calculating the maximum value of $\rho^{CM}_{Re}(\cdot, \cdot)$ and $\rho^{CM}_{NRe}(\cdot, \cdot)$, that is,

$$\rho^{CM}(\cdot, \cdot) = MAX(\rho^{CM}_{Re}(\cdot, \cdot), \rho^{CM}_{NRe}(\cdot, \cdot)). \qquad (21)$$

Fig. 3 illustrates the whole process of our watermark decoding technique.

4   Performance Analysis

Similar to the analysis discussed in [9], we shall try to estimate the probability of false negative (miss detection, failure to detect an existing watermark) and

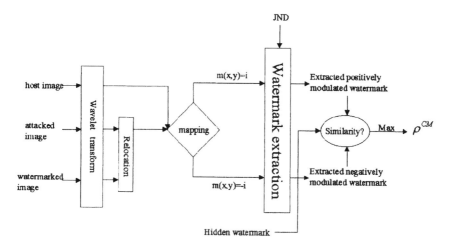

Fig. 3. The watermark decoding process of our cocktail watermarking scheme.

false positive (false alarm) as part of the proposed watermarking method. The probability of miss detection of our complementary watermarks is defined as

$$
\begin{aligned}
P_{fn}^{CM} &= P\{\rho(n_{pos}, n_{pos}^e) < T \ \& \ \rho(n_{neg}, n_{neg}^e) < T | a \ watermark\} \\
&= P\{\rho(n_{pos}, n_{pos}^e) < T | a \ watermark\} \cdot P\{\rho(n_{neg}, n_{neg}^e) < T | a \ watermark\} \\
&= P_{fn}^{pos} \cdot P_{fn}^{neg},
\end{aligned} \tag{22}
$$

where T is the threshold used to decide the existence of an extracted watermark. Index pos/neg denotes that the watermarks are embedded using positive/negative modulation rule and n/n^e represents the original/extracted watermark. Since the hidden watermark value is bipolar, the original and the extracted watermark values either have the same sign (i.e., $n_t(i)n_t^e(i) = 1$, $t \in \{pos, neg\}$) or have different signs (i.e., $n_t(i)n_t^e(i) = -1$, $t \in \{pos, neg\}$). It can be shown that $\sum n_t(i)n_t^e(i)$ belongs to the set $\{-L_M, -L_M+2, ..., L_M-2, L_M\}$ or $\sum n_t(i)n_t^e(i) = L_M - 2m$, where $m \in [0 \ \ L_M]$ and $t \in \{pos, neg\}$. Let p_1 be the probability of $n_t(i)n_t^e(i) = 1(t \in \{pos, neg\})$; it is equal to P_{nm}^A or P_{pm}^A, depending on the type of an attack. Then, we can derive P_{fn}^{pos} as

$$
\begin{aligned}
P_{fn}^{pos} &= P\{\rho(n_{pos}, n_{pos}^e) < T | a \ watermark\} \\
&= P\{\sum n_{pos}(i)n_{pos}^e(i) < L_M \cdot T | a \ watermark\} \\
&= \sum_{m=\lceil \frac{L_M(1-T)}{2} \rceil}^{L_M} P\{\sum n_{pos}(i)n_{pos}^e(i) = L_M - 2m | a \ watermark\} \\
&= \sum_{m=\lceil \frac{L_M(1-T)}{2} \rceil}^{L_M} \binom{L_M}{m} p_1^{L_M} (\frac{1-p_1}{p_1})^m.
\end{aligned} \tag{23}
$$

Similarly, P_{fn}^{neg} can be derived in the same way.

The individual derivation of P_{fn}^{pos} or P_{fn}^{neg} is similar to that of Kundur and Hatzinakos [9], but the result is extremely different due to (1) our p_1 is generated using a complementary modulation strategy and (2) two complementary watermarks are embedded by our method. It should be noted that the probability, p_1, in our scheme is lower bounded by 0.5. It can be expected that our false negative probability will be definitely smaller than those obtained by most methods with averaged $p_1 = 0.5$ (as discussed in Sec. 2.1). Besides, we would like to stress that it is not helpful in reducing false negative if multiple watermarks with the same role [3,14] are embedded. Finally, the analyzed result of false positive could also be derived as in [9].

The threshold T can be set automatically using Eq. (22) if a desired false negative probability is given. Under the condition that the watermark length L_M and the threshold T are fixed, our false negative probability is the lowest among the existing methods using random modulation. If we want to reduce the false negative probability, we have to decrease T but at the expense of increasing the false positive probability.

5   Experimental Results

A series of experiments was conducted to verify the effectiveness of the proposed method. First, a situation is indicated to show that bipolar watermark hiding is advantageous than real-style watermark hiding [12]. For example, Figs. 4(a) and (b) show a watermarked image and its distorted image by a brightness/contrast attack, respectively. Basically, the histogram of the watermarked image is significantly changed. By detecting real-style watermarks as shown in Fig. 4(c), the detector responses are plotted with respect to 1000 randomly generated watermarks. Unfortunately, the resultant detector responses correspond

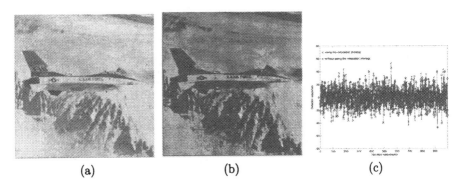

(a)   (b)   (c)

Fig. 4. Illustration of the drawback of noise-style watermark detection: (a) watermarked image; (b) brightness/contrast adjusted image; (c) the resultant detector responses correspond to the original watermarks 400 (using the relocation strategy) and 800 (without using the relocation strategy) are indistinguishable from others among 1000 random marks.

to the original watermarks 400 (using the relocation strategy) and 800 (without using the relocation strategy) are indistinguishable from others. However, if bipolar watermarks are detected, the detector response corresponding to the original mark is higher than 0.9. This example illustrates that even the signs of a extracted watermark are mostly preserved to be the same as those of the original watermark, their correlation calculated by Eq. 3 may have a chance to become small. In the following experiments, we will show the results of bipolar watermark detection.

The "Kids" image of size 128 × 128 shown in Fig. 5(a) is used as the host image. The length of a hidden watermark depends on the host image and the wavelet-based visual model. Here, its length was 1285. Using our complementary modulations, a total of 2570 wavelet coefficients needed to be modulated. The PSNR of the watermarked image (shown in Fig. 5(b)) was 40 dB. We used 20 different attacks to test the robustness of our watermarking scheme. The 20 attacked watermarked images are illustrated in Fig. 5(c0)~(c19). The detector responses, $\rho_{NRe}^{CM}(\cdot,\cdot)$, (without employing the relocation step) with respect to the 20 attacks are plotted in Fig. 6(a). The two curves clearly demonstrate the complementary effects. It is apparent that one watermark could be destroyed while the other one survived well. From the set of attacked watermarked images, it is not difficult to find that some attacks severely damaged the watermarked image, but the embedded watermarks could still be extracted with high detector response. It is obvious that the complementary modulation strategy enables at least one watermark to have high probability of survival under different kinds of attacks. Moreover, the detection results yielded from $\rho_{NRe}^{CM}(\cdot,\cdot)$ and $\rho_{Re}^{CM}(\cdot,\cdot)$ were also compared to identify the significance of relocation. Fig. 6(b) shows two sets of detector responses, one for detection with relocation and the other for detection without relocation. From Fig. 6(b), one can see that the asynchronous phenomena caused by attacks were compensated by the relocation strategy.

Our scheme was also compared with the methods proposed by Cox et al. [3] and Podilchuk and Zeng (IA-W) [14] under the same set of attacks. In order to make a fair comparison, the parameters used by Cox et al. [3] were adopted. The PSNR of their generated watermarked image was 24.25 dB. Podilchuk and Zeng's method [14] is image-adaptive and requires no extra parameter. The PSNR of their generated watermarked image was 30.75 dB. In our watermarking scheme and Podilchuk and Zeng's approach, 3-level wavelet transform was adopted for decomposing the Kids image. Among the three watermarked images generated, respectively, by Cox et al.'s method, Podilchuk and Zeng's method, and our method, our watermarked image had the highest PSNR. In order to make the comparison fair, the relocation step which would have made our approach even better was not used. A comparison of the detector responses with respect to the 20 attacks among the above three methods were illustrated in Fig. 6(c). From the comparison shown in Fig. 6(c), it is obvious that our scheme was superior to the other two, even though the watermark strength introduced by our method was the *weakest*.

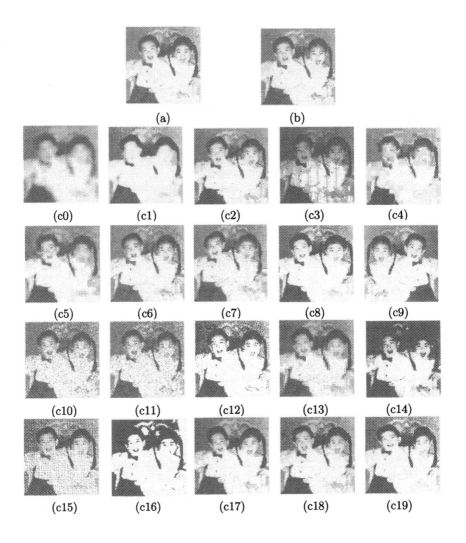

Fig. 5. (a) Host image; (b) watermarked image; (c0)∼(c19) attacked water-marked images ((c11)∼(c19) were generated using PhotoShop): (0) blurred (mask size 15 × 15); (1) median filtered (mask size 15 × 15); (2) rescaled; (3) histogram equalized; (4) JPEG compressed (with a quality factor of 1%); (5) SPIHT compressed (at a compression ratio of 64 : 1); (6) StirMark attacked (1 time with all default parameters); (7) StirMark attacked (5 times with all default parameters); (8) jitter attacked (5 pairs of columns were deleted/duplicated); (9) flipped; (10) sharpened (factor 85 of XV); (11) Gaussian noise added; (12) dithered; (13) mosaiced; (14) brightness/contrast adjusted; (15) texturized; (16) thresholded; (17) pinched; (18) rippled; (19) twirled.

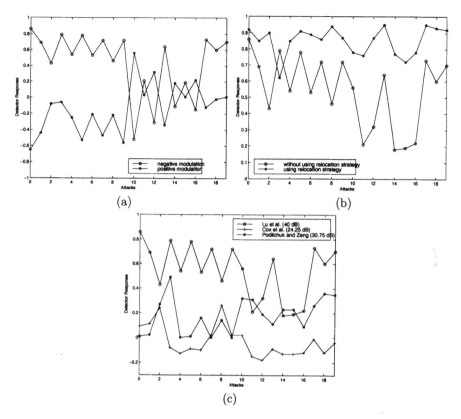

Fig. 6. Results obtained using our watermarking scheme (the maximum detector response was 1): (a) the detector responses (without using the relocation technique) under 20 different attacks; (b) a comparison of the detector responses with/without use of the relocation strategy; (c) a comparison between our method (PSNR 40dB), Podilchuk and Zeng's method (PSNR 30.75dB) [14], and Cox *et al.*'s method (PSNR 24.25dB) [3].

Moreover, we also take into account the Gaussian noise adding attack, which is recognized as the "balanced attack". The balanced attack means that it is usually balanced to within a close approximation by either increasingly or decreasingly updating the image pixels. With this property, one may argue that there is no benefit could be obtained from our technique. We will, thus, provide the following result to demonstrate that our scheme is able to resist the balanced attack like Gaussian noise adding. Gaussian noise with various amounts generated from Photoshop software were added into our watermarked image in Fig. 5(b). Visually, the image quality was severely degraded when the amount of Gaussian noise reached the level of 32. However, the detector response reflected in Fig. 7(a) turned out to be satisfactory, which stabilized at around 0.2 when the noise amount was over 32. In Fig. 7(b), we further show the uniqueness and the representative of the extracted watermark with the lowest detector response 0.2 (obtained after Gaussian noise adding with amount 32) among 10000 random marks.

From the above experimental results and our previous analysis, the advantage of our complementary modulation strategy is obvious. Although the methods of Cox *et al.* [3] and Podilchuk and Zeng [14] could easily employ similar redundancy by including multiple watermarks, this does not mean that their detection results will be as good as ours.

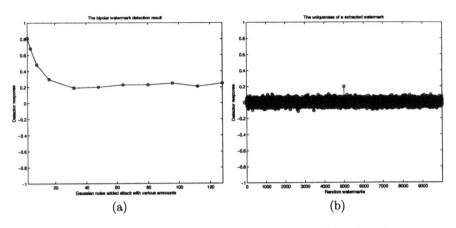

(a)   (b)

Fig. 7. Detection results after balanced attacks: (a) the detector responses obtained after Gaussian noise adding with various amounts (2,4,8,16,32,48,64,80,96,112,128); (b) the uniqueness of the extracted watermark (with lowest detector response in (a)) obtained after Gaussian noise adding with amount 32 among 10000 random marks.

6   Conclusion

A highly robust watermarking scheme, which can securely protect images, has been developed in this study. The proposed scheme has two characteristics: (1) embedding two complementary watermarks makes it difficult for attackers to destroy both of them; (2) a relocation step is included to deal with attacks that generate asynchronous distortions. Experimental results demonstrate that our watermarking scheme is extremely robust while still satisfying typical watermarking requirements. To the best of our knowledge, no other reports in the literature have presented techniques that can resist as many different attacks as our method can. Future work will focus on deadlock, registration, and capacity problems.

References

1. M. Antonini, M. Barlaud, P. Mathieu, and I. Daubechies, "Image Coding Using Wavelet Transform", *IEEE Trans. Image Processing*, Vol. 1, pp. 205-220, 1992.
2. M. Barni, F. Bartolini, V. Cappellini, and A. Piva, "Copyright Protection of Digital Images by Embedded Unperceivable Marks", *Image and Vision Computing*, Vol. 16, pp. 897-906, 1998.
3. I. J. Cox, J. Kilian, F. T. Leighton, and T. Shamoon, "Secure Spread Spectrum WaterMarking for Multimedia", *IEEE Trans. Image Processing*, Vol. 6, pp. 1673-1687, 1997.
4. S. Craver, N. Memon, B.-L. Yeo, and M. M. Yeung, "Resolving Rightful Ownerships with Invisible Watermarking Techniques: Limitations, Attacks, and Implications", *IEEE Journal on Selected Areas in Communications*, Vol. 16, pp. 573-586, 1998.
5. J. F. Delaigle, C. De Vleeschouwer, and B. Macq, "Watermarking Algorithms based on a Human Visual Model", *Signal Processing*, Vol. 66, pp. 319-336, 1998.
6. J. Fridrich, "Combining Low-frequency and Spread Spectrum Watermarking", *Proc. SPIE Int. Symposium on Optical Science, Engineering, and Instrumentation*, 1998.
7. F. Hartung and B. Girod, "Watermarking of uncompressed and compressed Video", *Signal Processing*, Vol. 66, pp. 283-302, 1998.
8. F. Hartung, J. K. Su, and B. Girod, "Spread Spectrum Watermarking: Malicious Attacks and Counterattacks", *Proc. SPIE: Security and Watermarking of Multimedia Contents*, Vol. 3657, 1999.
9. D. Kundur and D. Hatzinakos, "Digital Watermarking Using Multiresolution Wavelet Decomposition", *Proc. IEEE Int. Conf. Acoustics, Speech and Signal Processing*, Vol. 5, pp. 2969-2972, 1998.
10. M. Kutter, F. Jordan, and F. Bossen, "Digital Signature of Color Images using Amplitude Modulation", *Journal of Electronic Imaging*, Vol. 7, pp. 326-332, 1998.
11. C. S. Lu, S. K. Huang, C. J. Sze, and H. Y. Mark Liao, "A New Watermarking Technique for Multimedia Protection", to appear in *Multimedia Image and Video Processing*, eds. L. Guan, S. Y. Kung, and J. Larsen, CRC Press Inc.
12. C. S. Lu, H. Y. Mark Liao, S. K. Huang, and C. J. Sze, "Cocktail Watermarking on Images", *to appear in 3rd International Workshop on Information Hiding*, Dresden, Germany, Sept. 29-Oct. 1, 1999.
13. F. Petitcolas, R. J. Anderson, and M. G. Kuhn, "Attacks on Copyright Marking Systems", *Second Workshop on Information Hiding*, USA, pp. 218-238, 1998.
14. C. I. Podilchuk and W. Zeng, "Image-Adaptive Watermarking Using Visual Models", *IEEE Journal on Selected Areas in Communications*, Vol. 16, pp. 525-539, 1998.
15. J. J. K. Ruanaidh and T. Pun, "Rotation, Scale, and Translation Invariant Spread Spectrum Digital Image Watermarking", *Signal Processing*, Vol. 66, pp. 303-318, 1998.
16. M. D. Swanson, B. Zhu, and A. H. Tewfik, "Multiresolution Scene-Based Video Watermarking Using Perceptual Models", *IEEE Journal on Selected Areas in Communications*, Vol. 16, pp. 540-550, 1998.
17. A. B. Watson, G. Y. Yang, J. A. Solomon, and J. Villasenor, "Visibility of Wavelet Quantization Noise", *IEEE Trans. Image Processing*, Vol. 6, pp. 1164-1175, 1997.

Region-Based Watermarking for Images

Gareth Brisbane†, Rei Safavi-Naini†, and Philip Ogunbona‡

†School of IT and CS
University of Wollongong,
Northfields Ave.
Wollongong 2522, Australia

‡Digital Technology Research Lab
Motorola Australian Research Centre

E-mail: {gareth, rei}@uow.edu.au, pogunbon@arc.corp.mot.com

Abstract. There is a spate of research activities investigating digital image watermarking schemes in an effort to control the problem of illegal proliferation of creative digital data across the Internet. Several schemes, with their merits and demerits, have been proposed to date. However, many of them rely on the presentation of a test image in an appropriate format before the detection can proceed. In most cases this is achieved through some form of pre-processing. This paper proposes a technique, using image region segmentation, which alleviates many of the problems associated with pre-processing and thus allowing the detection process to be more independent as well as reducing the need for human intervention.

1. Introduction

Digital information is now readily available due to advances in the compression, storage and communications technologies. The amount of digital information that can be found on the Internet and the popularity of the Internet corroborate this observation. Unfortunately, the protection of this information, especially in circumstances where the owners hope to generate revenue through controlled dissemination, is yet to be standardized. There are at least two consequences of the status quo. First, digital information that is already available is being illegally re-distributed and thereby robbing the legal owners of deserved revenue. Second, this situation discourages content generators from sharing their work with the wider community. A solution to the latter case is the development of proprietary techniques to insert and verify digital watermarks.

Several techniques have been proposed in the literature to insert watermarks in digital images. These techniques provide various degrees of robustness. Watermarking applications include, *copyright protection, authentication, embedded and hidden annotations,* and *secure and invisible communications.* Watermarking systems that are intended for copyright protection are likely to be subjected to a wide range of attacks that are aimed at removing the watermark, or simply making it ineffective and so a very high degree of robustness is required. Watermarking systems for authentication belong to the *fragile* class of schemes where the slightest change to the image completely destroys the mark and hence the authenticity of the image. Finally watermarking for embedding annotations and secure and invisible

M. Mambo, Y. Zheng (Eds.): ISW'99, LNCS 1729, pp. 154-166, 1999.
© Springer-Verlag Berlin Heidelberg 1999

communication require resistance against moderate level of modification due to routine image manipulations including digital to analog conversion and vice versa.

In this paper we are concerned with a group of attacks on a watermarked image that result in the loss of synchronization and hence failure of the watermark verification process. By synchronization we mean correct correspondence between pixels: that is the ith pixel in the original image remains the ith pixel in the image under verification test. The loss of synchronization may be due to intentional attacks, such as deleting a row or a column of pixels, or unintentional causes such as misalignment during the scanning of the image.

The problems of misalignment and loss of synchronization are discussed in some detail in [1], where the authors reviewed several techniques for overcoming these problems. The solution proposed is to repeatedly test shifted versions of the verification signal, searching for a high peak to indicate correlation. However, the application of this scheme is only discussed in terms of binary insertion and detection, which places in the class of fragile watermark. Also, the notion of continuously testing different signals to search for matches is unappealing.

Another attempt to produce a region-based watermarking algorithm was by Schneider and Chang [2], where individual objects in an image were targeted for watermarking. However, this was a procedure that required manual selection of image regions for both insertion and detection. Our objective is to use image segments, which may not necessarily correspond to an object, but are generated by a simple segmentation algorithm.

Segmentation methods can be as simple as partitioning an image into tiles, or through the use of image features that capture the quantitative description of a homogeneity criterion. Region growing, clustering and watershed algorithms are some examples of feature based segmentation. In general, the pixels within each segmented region will be similar according to some criteria.

Simple image partitioning or tiling is commonly used for increasing efficiency in watermarking systems. The problem of efficiency arises when image sizes do not lend themselves to operation by a fast transformation algorithm. An example can be found in the case of the DCT where the fast radix-2 algorithm is well known. Images with sizes that are not multiples of 2^N, where N is an integer, have to be tiled into sub-images of sizes amenable to the fast algorithm.

We propose segmentation for *efficiency, robustness against synchronization loss* and *localization*. Using segments and breaking the verification process into smaller chunks results in more efficient computation. This is especially true when a hardware implementation of the watermarking process is required. Using robust segmentation algorithms and inserting the watermark in selected secret segments will make attacker's attempts to cause synchronization loss much more difficult. Lastly, it may be required to localize the regions of the image where possible changes have to be detected. This has the advantage of potentially allowing a form of group watermarking, where each member can place a watermark in a different component of the image.

Segmenting an image into regions and inserting the watermark only in the desired regions allows various levels of protection for different part of an image. However the use of segmentation requires careful design of the system as it could adversely effect synchronization loss because of the introduction of many regions that need to be individually synchronized. The segmentation process during verification process must recover exactly the same regions that have been used during the insertion process. In other words, segments must remain *invariant* after the watermark insertion process.

For example, in [3], the authors proposed the use of an invariant domain to store the watermark information. An *invariant domain* is an area of the image that will always be perfectly reconstructed irrespective of the changes introduced to the image. In [3], the domain proposed is invariant to rotation, scaling and translation; such attacks will not have any effect on the retrieval of the watermark. Fleet and Heeger [4] introduced a watermarking scheme that is rotation and scale invariant. The insertion of the watermark was in the Yellow-Blue colour space of the image and invariance was achieved by modifying the amplitude of regularly spaced sine waves inserted in this domain. At a later stage, these waves can be recovered using the same technique and the image can be realigned and re-scaled to its original dimensions. Cox et al use the whole of frequency range to provide an invisible yet robust domain to insert the watermark [5]. While this domain is robust, it is not invariant, as changes in the spatial domain via rotation for example, will substantially affect the reconstruction of the domain.

We propose a simple magnitude based ordering of pixels before the insertion process; the ordering is necessarily undone before verification takes place. The reasoning behind the ordering is that this process relies on the relative magnitude of data, in this case pixel values, and so is independent of the orientation of a segment during the re-scanning of the image. We note that in each of the above-mentioned techniques, the domain of insertion is publicly accessible, although the location of the watermark itself may be preserved by use of the owner's secret information. Using segmentation allows us to hide the domain of insertion if required.

The rest of this paper is organized as follows. In section 2 we develop the intuition behind our approach and provide justification for its operation. In section 3 we give the outline of an implementation using a simple segmentation method in conjunction with the scheme proposed by Cox et al. Our experimental output is documented in section 4. Section 5 then provides some concluding remarks.

2. Segmentation in Watermarking

Watermarking consists of two complementary processes: *watermark insertion* and *watermark verification*. During the insertion process a watermark signal, w, is embedded into a host object I and produces a watermarked object $I(w)$. The watermarked object might undergo a range of intentional and unintentional modifications. The aim of the verification process is to produce evidence in support of the existence of the watermark. The verification algorithm produces a *true* result by recovering the watermark, possibly in a distorted and weakened form, but with appreciable signal strength that supports existence of the mark with a high probability. The main challenge in robust watermarking is to ensure that the system produces correct results. In other words, verification should produce *true* for all valid marks, and *false* for all fraudulent objects, if the object has been subjected to a range of intentional and unintentional modifications. It is important to note that because of various unintentional modifications, such as filtering and compression, certain degree of robustness is necessary for *all* watermarking applications. Furthermore, synchronization is lost if an object is not perfectly aligned and may result in the failure of the verification process. To illustrate this problem, consider the case of computing a transform (e.g. DCT) of a tiled image in order to extract the watermark signal. The image needs to be properly aligned so that the same tiles used in the insertion process can be recovered. If the watermarked image were visually imperceptibly skewed or rotated the verifier requires knowledge of this process to be able to perform the verification process accurately. This problem remains even when the whole image (rather than tiles) is used during the insertion process.

Segmentation-based watermarking attempts to provide the required invariance property when verifying a watermark in the scenarios described above. Region segmentation based on textural properties of the image can be robust to affine transformations if appropriate descriptors are used. It is also *robust* against minor changes aimed at removing the strict correspondence between the original image and its watermarked form. Such image modifications might include row (or a column) removal, substitution of a row (or a column) with the average of adjacent ones, line insertion and resizing.

2.1 The System

We assume an original image, I, consisting of an array $N \times M$ of pixels which could be grayscale or colour. Let I_s denote the sth segment of the image, where each segment consists of a subset of pixels forming an arbitrarily shaped area of the image and obtained through a possibly human-assisted algorithm. Our definition of a segment is assumed broad enough to subsume the case of tiling and the case where a human user selects areas of an image as segments. Let S be the total number of segments. Each segment I_s from the image can be ordered into a sequence, x_s, of pixels, with the ith component denoted as $x_s(i)$. X denotes the set of all such segments. The watermarking process usually requires some input called *key* and denoted by k which is only accessible to the owner of the object or in general the person who inserts the mark. We use K to denote the set of possible keys and $f_K(x)$

to denote a key-dependent transformation on an input segment, x_s; $f_K(x)$ is for example a watermark insertion algorithm. A watermark detection algorithm, $g_K(x, y')$, produces an estimate of the likelihood that an image segment represented by the sequence y' contains the watermark. In other words, $y' = f_k(x)$. For an image with segments y_s, the likelihood of being derived through watermarking from the image segment set $\{x_s \in X; s = 1, \cdots, S\}$ is calculated as $W = \sum_{s=1}^{S} g_k(x_s y_s)$.

2. 2 Producing Invariance in Segments

Consider the simplest case where the number of segments, S , is equal to one. In other words the whole image is used for insertion and detection. One drawback of the whole image as one segment is computational and implementation inefficiencies. The system is also prone to loss of synchronization if the sequence, x_s (in this case the whole image), derived from the original image does not bear a one-to-one correspondence to the sequence, y_s, derived from the watermarked image. The problem faced when $S > 1$ is no less, because any inaccuracy in determining the segment boundaries will have a cumulative effect when more segments are considered.

The detection function $g_k(x, y)$ can be seen as a correlator that provides comparison between the two vectors and ensures that the differences between the two are predictable by k. If the correspondence between segments of I and $I(w)$ is not accurate the correlation measure fails. Note that the loss of segment synchronization can occur if a simple transformation such as cropping or adding an extra column to the image is performed. It is therefore necessary to reverse the operation before any comparison between two sequences are made. In the case of cropping, the remaining portion of the picture must be replaced in the original position, so that the correct neighbourhood can be reconstructed for detection (as indicated in [5]). If a column has been added, then it must be removed to regain synchronization. Finally for segmentation to work properly, we must ensure that each sequence x_s relates to y_s on a point-to-point basis. This is achieved through the ordering mechanism described before.

It is desirable that it be able to detect the change in segment boundaries and recognize the need to realign the segments. This requirement is particularly important with schemes that depend on inserting a watermark at certain positions in the image, e.g. "Patchwork"[6]. Here we have a fixed point somewhere within I , the reference image. By mapping the watermarked image $I(w)$ onto I we can recover the differences and indicate the presence of the watermark. If the reference point is no longer valid, even if it has only been slightly tampered with, then the comparison cannot be correctly made.

To counter these problems, some attempts have been made to provide an invariant domain of operation to allow $g_k(x, y)$ to work correctly, irrespective of changes made to the image [3]. In their work [3], Ó Ruanaidh and Pun achieved invariance by transforming the image through the Fast Fourier Transform, then through the Log Polar Transform and finally through another Fast Fourier transform. The end result of this process is a domain that is invariant to rotations, scaling and translations. However, cropping is an effective attacking tool, which does not degrade the appearance of the original image, although it does reduce the amount of original information. Another technique attempts to introduce a small mask of sine waves across the image [4], which can be retrieved at a later stage to indicate how the image has been stretched or rotated. Unfortunately, these are produced so as to be public knowledge and so can be filtered out by a clever attacker, or modified to add credence to the attacker's version of the image.

The foregoing discussions demonstrate the need for an invariant segmentation technique which does not rely on a reference pixel which can be removed from an attackers duplicate, nor on the order of the pixels, which can be changed at whim through rotations or other symmetrical transformations. We propose a technique to redress this problem by segmenting the image. Our method does not regard the image as a segment since this would require the presence of all original pixels during verification. Furthermore, we do not simply tile the image, as this requires direct synchronization of the tiles for verification. In the implementation presented in this paper, we use simple region growing segmentation method that relies on pixel properties to define regions. We quickly point out that any image segmentation algorithm that uses invariant local properties to describe a pixel is applicable. Using such an approach, we expect to achieve high level of robustness against loss of synchronization. We can also obviate the need for having the pixels in the same order by enforcing a fixed predetermined order on the pixels before inserting the mark. Thus, we can effectively provide efficient watermarking functions, $f_K(x)$ and $g_K(x, y)$, which are robust against loss of synchronization. In summary we propose two methods which can provide an invariant domain:

1. Use a segmentation method based on some similarity criterion to partition the image into segments that only rely on pixel properties, and store enough information about the segments to enable exact recovery of segments during the verification phase. Techniques that can be employed to achieve the region segmentation include among others, k-means clustering, fuzzy c-means, region growing and watershed. This was performed in a manual fashion in [2], where the content of the image was targeted for image authentication.
2. Impose a fixed order, for example ascending order of magnitude of pixel values, on pixels within a segment before inserting the mark. The effect of the imposed ordering is to provide robustness (invariance) against rotation or misalignment in a single segment.

3. Feature-Based Segmentation

A simple region growing algorithm has been used in our tests. The feature-based segmentation utilizes each of the colours, red, green and blue in turn. It moves across the image, selecting key pixels. In our tests, each key pixel was 10 pixels apart from its nearest neighbour in all four directions. From each of these pixels, every adjacent pixel is added to the region if their difference is minor. The strict definition of a minor difference is left to the discretion of the user, but for our tests we considered two pixels to be similar if the colour being used is within 20 grey levels (on a scale of 256 grey levels). We also took the minimum size for a region to be 3,000 pixels in total area so as to prevent the watermark relying on too small an area for data hiding.

Segments are required to have a minimum size to be considered for watermarking, otherwise they are merged with neighbouring regions. Therefore, arbitrarily small regions will not be marked as they are of less importance to the image than the larger regions. In this way, an image can be broken up into many different segments for watermarking with the important property that, if the image is cropped or rotated, we should still be able to recover the same segments. If the image is scaled, we then recover segments that are proportionately smaller. The value of this is that if an attacker decides to extract only the important regions of the image then the detection process will be able to identify the same segments as with our original image and use $g_k(x, y)$ in matching segments. In this way, we have achieved our goal of invariant segmentation and watermark insertion.

Because of the way we convert the two-dimensional regions into one-dimensional arrays, the scan line plays an important part in determining how the watermark is spread across the region. Suppose we extract S segments (or regions) from an image. Taking our sequence x_s, we then obtain $y_s = f_K(x_s)$. Finally we recompose our image by replacing the segments in their original positions. The proposed method is able to achieve invariance even with rotated images.

Detection follows a similar process where we generate segments in the same fashion as for insertion. To allow correct verification we must be able to correctly recover the segments. However because of the insertion of the watermark, we cannot expect exact recovery unless some key values from the original segmentation process is stored. The amount of stored data is an important efficiency parameter of the system. An efficient system should require relatively small size of essential segmentation data, while a naïve system might require all segment outlines to be stored.

4. Experimental Results

In order to verify the feasibility of the proposed techniques, the test image, "Peppers", was segmented into regions using the simple region growing techniques described earlier. As mentioned above, we put the threshold for related pixels at 20 shades and regard all regions under 3,000 pixels as being too small. Simulation of watermark insertion based on Cox's method [5] was performed. For each of the segments generated, the watermark amplitude was set to 0.1 and the lowest 1,000 frequencies of the transformed segments were modified. We note that in our simulation, we did not use the first post-processing suggested in [5], viz., mean subtraction. Rather, we employed only the second type of post-processing operation, by allowing for tolerance levels in the change in the transformation coefficient values.

The actual watermarking pattern is based on the experiments mentioned in [5]. It is important to note that the detection scores we calculated were not quite as impressive as those demonstrated in that paper. For using an entire image with a thousand frequencies, the benchmark score was 32.0, while our results yielded 23.55. Likewise, after using JPEG compression at 90% level, they still retain enough watermark content to give 22.8, while for us, we achieve only 3.97. With the current results, we cannot claim robustness to distortions introduced by compression. There are, however, ways in which we can improve the technique. For example, the choice of a transform more amenable to the structure of the reordered pixels. The DCT transform used by Cox may not be the most appropriate transform to use, given that the sorting operation will create a domain which represents a ramp function.

4.1 Region Growing

The detection process used the same regions that were generated in the original image, so there would be no discrepancies between the regions. Of course, the ultimate objective is to have each image independently generate its own region set and automatically compare the closest matches in the detection process.

Figure 4.1.1 shows the original "Peppers" image prior to the watermarking process. Figure 4.1.2 illustrates the five regions that were constructed by the clustering process. Figure 4.1.3 gives us the watermarked image, achieved after applying Cox's method to the five regions found in Figure 4.1.2. Figure 4.1.4 attempts to show the level of confidence across the watermarked image. It shows a scale from dark to light, where black indicates no contribution to the watermarking score (due to the lack of an appropriate region), grey indicates no contribution the watermarking score, while white represents a strong confidence level in the presence of the watermark.

The detection process scored the inserted watermark with 35.10. This value easily demonstrates the presence of the watermark. The Peak Signal to Noise Ratio (PSNR) was used as the image fidelity measure. The PSNR of image shown in Figure 4.1.3 relative to that in Figure 4.1.1 was 38.80dB and showed little visible variation between the two images.

However, the need to produce consistent and invariant regions in the image causes some difficulty. Here in this experiment, we stored the image map which shows how the regions are spread across the original image (as shown in Figure 4.1.2) in Run Length Encoding (RLE) form as a simple and effective technique for reducing the amount of stored information required. The image map by itself would have taken up 65,536 bytes but after the encoding, the size was reduced to 10,224 bytes. This was necessary, as using region-growing segmentation on the watermarked image produces regions that are too dissimilar to reliably detect the watermark. The reason for this is that the watermark insertion process modifies the pixel values. Thus, the regions which are detected using the same process vary from the originally generated regions.

Figure 4.1.1 Original Peppers Image

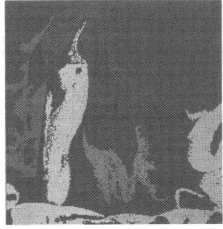

Figure 4.1.2 Peppers image showing the formed feature-based regions

Figure 4.1.3 Watermarked Peppers image using formed feature-based regions

Figure 4.1.4 Peppers image illustrating the confidence level of the recovered watermark in figure 4.1.3

4.2 K-Means

In these experiments the components of the feature vector used by the k-means algorithm are simply the pixel under consideration and its four-neigbour pixels. Effectively, we have not taken advantage of the availability of colour information. Five regions have been chosen in these experiments and the result is presented Figure 4.2.1.

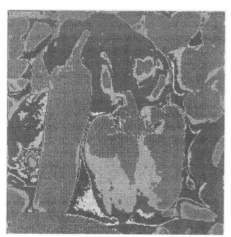

Figure 4.2.1 Peppers image showing the 5 k-means generated regions

Figure 4.2.2 Watermarked Peppers image using k-means generated regions

Figure 4.2.3 Peppers image illustrating the confidence level of the recovered watermark in Figure 4.2.2

The codebook that generated the regions were stoored using RLE. The detection score for Figure 4.2.2 is 36.67 and has a PSNR of 36.16dB. Although the output image is noisier, it is important to note that the entire image now has a watermark inserted as opposed to Figure 4.1.3, which has only had the valid regions watermarked.

Number of regions	Watermark detection score	PSNR (db)
3	33.39	37.32
4	34.18	37.00
5	35.04	37.00
6	37.00	35.12
7	36.61	34.44
8	39.49	35.35

Table 4.2.1 Average of 10 watermarked images seeded with consecutive numbers

The results presented in Table 4.2.1 indicate that the detection scores improve as we segment into more regions. We do expect this at the cost of robustness because we have given ourselves more locations to store information at the cost of the spread of the watermark. In general though, the benefit of providing more regions is that the locality of the watermark becomes stronger. This means that if an image is cropped, a greater portion of the watermark remains in that area, than if it had been spread over the entire image. The cost to the PSNR of using more regions is minimal.

4.3 Sorting

The results obtained when sorting was incorporated with k-means or region-growing based segmentation were not encouraging because the insertion process via the DCT disrupted the order of reconstructed pixels. This problem is being investigated further with the view to selecting a more appropriate transform.

5. Conclusion

In this paper we proposed segmentation and pixel ordering as a general method of implementing robust watermarking systems. Segmentation can provide higher robustness and efficiency, and also a way of selectively watermarking parts of an image. The latter application is important if different parts of an image have different semantic significance and so require higher level of protection. However introducing segmentation can result in more susceptibility to loss synchronization. We noted that correct detection requires correct recovery of the segments during verification phase. Towards this goal some key information from the original segmentation process must be preserved and used in the verification phase. At present, we require an original "image-map" to recover the watermark. In the region growing segmentation method given in Section 3 the size of this key information was rather large, that is equal to the number of pixels in the image (65,536 bytes) or the RLE of the regions in the image (10,224 bytes). In the second clustering method, given in section [2], this could be reduced to (number of regions * 9 * 8 bytes) which is a dramatic reduction. The reduction is because of the use of a clustering algorithm that allows the recovery the segments from the stored values.

Although finding the optimal clustering algorithm requires further research, the approach is potentially very powerful. It means that we are no longer bound by the dimensions of an image or by the relative position of the components within the image. Watermark verification can then be done by an automated process without the need for human pre-processing to guarantee good results. If necessary, the regions could be selected by hand, or pre-determined such as with object recognition or by provision of the digital media itself, to achieve good results.

The use of an invariant technique would establish simple verification of any image to provide identification and/or verification of ownership as it is independent of the modifications of the image. It is important to note that the underlying watermarking algorithm is of no key significance. This is because most contemporary watermarking algorithms make no distinction between being given an image or given just a segment to work with. In this way, an invariant technique that provides these segments will allow the algorithm to work as effectively as if it were working on the entire image.

Further work involves testing these relationships and verifying that segment comparisons are effective and efficient. We also need to establish other means to use sorting to resist the changes introduced by the DCT. There are many sides to this problem that need investigation, such as the way clustering is performed and also the level to which segments can be easily verified outside of their original contexts.

Acknowledgments:

This research work is partly funded by Motorola Australian Research Centre, through the Australian Government SPIRT grant scheme.

References:

[1] A. Z. Tirkel, C. F. Osborne, and T. E. Hall, "Image and watermark registration", Signal Processing, vol 66, pp373-383, 1998.
[2] M. Schneider and Shih-Fu Chang, "A Robust Content Based Digital Signature for Image Authentication", Proceedings of the 1996 International Conference in Image Processing, pp 227-230, 1997.

[3] J. O. Ruanaidh, W. Dowling, and F. Boland, "Phase Watermarking of Digital Images", Proceedings of the 1996 International Conference on Image Processing, vol. 3, pp 239-242, 1996.

[4] D. Fleet, and D. Heeger, "Embedding Invisible Information in Color Images", Proceedings of the 1997 International Conference on Image Processing, pp 532-534, 1997.

[5] I. Cox, J. Kilian, T. Leighton, and T. Shamoon "A Secure, Robust Watermark for Multimedia", NEC Research Institute, Technical Report 95-10, 1995.

[6] W. Bender, D. Gruhl, N. Morimoto, and A. Lu, "Techniques for data hiding", IBM Systems Journal, vol. 35, pp313-336, 1996.

[7] A. Gersho and R. M. Gray, *Vector Quantization and Signal Compression*, Kluwer Academic Publishers Group, 1991.

Digital Watermarking Robust Against JPEG Compression

Hye-Joo Lee[1], Ji-Hwan Park[1], and Yuliang Zheng[2]

[1] Department of Computer Science,
PuKyong National University, the Republic of Korea
leehj@woongbi.pknu.ac.kr,
jhpark@dolphin.pknu.ac.rk
[2] School of Comp. & Info. Tech., Monash University, Australia
yuliang@pscit.monash.edu.au

Abstract. Digital watermarking has been considered as an important technique to protect the copyright of digital content. For a digital watermarking method to be effective, it is essential that a watermark embedded in a still or moving image resists against various attacks ranging from compression, filtering to cropping. As JPEG is a dominant still image compression standard for Internet applications, digital watermarking methods that are robust against the JPEG compression are especially useful. Most digital watermarking methods proposed so far work by modulating pixels/coefficients without considering the quality level of JPEG, which renders watermarks readily removable. In this paper, we propose a new method that actively uses the JPEG quality level as a parameter in embedding a watermark into an image. A useful feature of the new method is that the watermark can be extracted even when the image is compressed using JPEG.

1 Introduction

The tremendous development in data compression methods has resulted in the widespread use of digital data such as image, audio and video in every corner of our daily life. Digital data are easy to distribute and duplicate. This gives rise to a serious problem in illegal copying. While encryption is essential to the provision of confidentiality, the same technology does not represent an ideal solution to copyright protection, simply because any user who possesses a decryption key may (re-)distribute decrypted digital images as he or she wishes. This indicates the necessity of embedding information on the rightful owner into an image in such a way that the information and the image cannot be easily separated. To achieve this goal, researchers have proposed to use so-called digital watermarking [1,2]. The most important requirement of digital watermarking is that embedded watermarks are robust against compression, filtering, cropping, geometric transformation and other attacks. For images that are published on the World-Wide Web (WWW), robustness of watermarks against the JPEG compression standard is particularly important.

M. Mambo, Y. Zheng (Eds.): ISW'99, LNCS 1729, pp. 167–177, 1999.

In this paper, we propose a method that constructs a watermark by using the quality level of JPEG as a parameter. As a result we obtain a watermarking method that is robust against the JPEG compression. We describe an overview of digital watermarking in Section 2. This is followed by a detailed description of our proposed method in Section 3. A number of simulation results are provided in Section 4 to verify the effectiveness of the proposed method.

2 Digital Watermarking Method

Digital watermarking consists of a pair of matching procedures, one for embedding a watermark into a still or moving image and the other for detecting/extracting the watermark. A number of factors have to be considered while embedding a watermark. These factors include the structure of a watermark, locations where the watermark is embedded, and the level of change in the quality of the image introduced by digital watermarking.

The structure of a watermark generally falls into one of two types. The first type is essentially a random binary sequence which is composed of either 0 and 1 or -1 and 1 [3,4,5,6]. The second type is a random real number that is distributed according to $N(0,1)$[7]. With a random binary sequence, one can apply the sequence in the extraction of an embedded watermark as well as the detection of the presence of the watermark. A disadvantage of the use of a random binary sequence is that the watermark is vulnerable to such attacks as removal and collusion. In comparison, with a random real number, one cannot extract the original image. Nevertheless, this method has the advantage of being more robust against removal and collusion attacks, primarily due to the fact that even though an attacker may have some knowledge on the locations of a watermark, he or she has far greater difficulties in identifying the exact watermark. Random real numbers are being used by researchers more often than random binary numbers recently.

Locations for embedding watermark are often determined by using random sequences together with human visual system(HVS)[8,9]. For instance, the Podilchuk-Zeng method[8] have utilized the visual model developed by Watson[10] for the JPEG compression. More specifically, the authors have used the frequency sensitivity portion of the model to embed a watermark. In addition, Dittmann[9] et al have developed a robust video watermarking method based on a combination of an error correction code, the Zhao-Koch method[11] in the frequency-domain and the Fridrich method[12] in the spatial-domain. The method proposed in this paper is similar to the above two methods in the use of the HVS. However, as our method embeds copyright information in the spatial domain, it differs from the Podilchuck-Zeng method that embeds in the frequency-domain using DCT or wavelet transform. Compared with the method by Dittmann et al, their method doesn't consider the re-compression of a watermarked video image.

With a method entirely relying on a secure random sequence to determine locations for embedding a watermark, the quality of a watermarked image degrades since it does not fully consider the feature of the image. In contrast, by using HVS in deciding locations, it is possible to make the change of image perceptually unrecognizable. Nevertheless an attacker who is familiar with HVS may still be able to estimate the location of the watermark. For this reason, the use of HVS actually decreases the secrecy with respect to the watermark's locations even though it meets the requirement of invisibility.

In view of these observations, we argue that digital watermarking should satisfy the following conditions:

1) If the structure of a watermark is a binary random sequence, the magnitude of the watermark has to be different in all pixels.
2) If locations to embed the watermark are random, the magnitude of the watermark has to be dependent on the image.
3) If it employs HVS, locations to embed the watermark have to be random.

In the following section, we describe our proposed method that converts a text on copyright information into binary random sequences. To satisfy the above conditions, the embedding locations and changes in pixels are determined by using the HVS.

3 Digital Watermarking Using Difference

The JPEG algorithm is one of the loss compression methods for still images. We consider a situation where a watermarked image will be compressed using the JPEG prior to its publication on WWW. Figure 1 illustrates the block diagram of the JPEG algorithm.

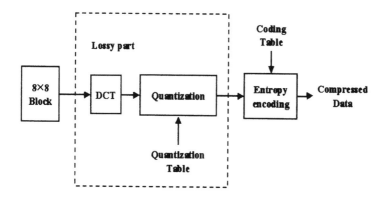

Fig. 1. The JPEG compression

As illustrated in Figure 1, the loss of information is due to quantization. When the JPEG algorithm is performed, *the quality level* is used to construct a quantization table. In general, the higher the quality level, the lower the compression ratio. Now consider a case where users uses only JPEG for compression. When the owner of an image expects to achieve robustness at least for the quality level q_1, he or she can construct a watermark, and embeds it into the original image by using q_1 as a parameter. We will show how this can be done. In addition we will show that it is possible for the owner to extract the watermark from the watermarked image when the watermarked image is compressed with JPEG at a higher quality level $q_2(> q_1)$.

3.1 Preprocessing

To start the method we construct a watermark as follows. We calculate the differences between the original image and a reconstructed image after compression with q_1, and the visual component from the original. Each parameter is defined as follows.

i, j : denote positions of pixels on the original image of size $N \times M$, $0 \leq i < N$, $0 \leq j < M$

h, v : denote indexes of a 8×8 block which is partitioned from the original image, $0 \leq h < N/8$, $0 \leq v < M/8$

k, l : denote positions within one block, $0 \leq k < 8$, $0 \leq l < 8$.

First, an image I of size $N \times M$ is partitioned into 8×8 blocks. Subsequently, DCT transform is applied to each block and all DCT coefficients are quantized by quantization factors obtained from q_1. Note that the quality level q_1 acts also as an indicator of robustness against the JPEG compression. For the reconstructed image, the difference $D_{h,v}$ of all blocks is calculated using Eqn.(1).

$$D_{h,v} = \{d_{k,l} | x_{k,l} - x'_{k,l}\}, \tag{1}$$

where $x_{k,l}$ and $x'_{k,l}$ denote the value of a pixel in the original image and the value of a reconstructed pixel with respect to q_1. The difference $d_{k,l}$ is used to decide the magnitude of a watermark to be embedded, and this magnitude is then used to obtain the watermark patterns.

The next step in preprocessing is to compute visual components that are dependent on the image. Centering the pixel value $x_{i,j}$ within a $t \times t$ window, the average difference of brightness $\hat{x}_{i,j}$ between the centering pixel and its neighboring pixels is calculated using Eqn.(2).

$$\hat{x}_{i,j} = \frac{1}{t^2 - 1} \left\{ \sum_{s_1=i-t}^{i+t} \sum_{s_2=j-t}^{j+t} |x_{s_1,s_2} - x_{i,j}| \right\} \tag{2}$$

The average brightness $b_{i,j}$ is then calculated for the neighboring pixels using Eqn.(3)

$$b_{i,j} = \frac{1}{t^2 - 1} \left\{ \sum_{s_1=i-t}^{i+t} \sum_{s_2=j-t}^{j+t} x_{s_1,s_2} - x_{i,j} \right\}. \tag{3}$$

From the average brightness $b_{i,j}$, a relative intensity R is obtained via Eqn.(4).

$$R = \left[\frac{x_{i,j}}{b_{i,j}} \right] = [r_{i,j}] \tag{4}$$

It is difficult to perceive the change by human eyes, as the change takes place in the vicinity of edge. Therefore, we can utilize the following as visual components.

$$V_{h,v} = \{v_{k,l}| \log(r_{k,l} \cdot \hat{x}_{k,l})\}. \tag{5}$$

As shown in Eqn.(5), the visual components are obtained by multiplying $\hat{x}_{i,j}$ by $r_{i,j}$ and scaling the resulting value with log. The values $d_{k,l}$ and $v_{k,l}$, which are dependent on the image, are used to produce a watermark that satisfies the conditions 1) and 2) as mentioned earlier.

3.2   Construction of a Watermark

Before locations to embed a watermark are determined, we have to compute the magnitude of the watermark from the values of $d_{k,l}$ and $v_{k,l}$. The values of the watermark are restricted by the difference value $d_{k,l}$ through a modular operation. A pattern $U_{h,v}$ can be obtained from the product of $D_{h,v}$ and $V_{h,v}$. Note that the pattern is actually a matrix of size 8×8 (see Eqn.(6)).

$$U_{h,v} = (D_{h,v} \times V_{h,v}) \bmod D_{h,v}, \tag{6}$$

It should be pointed out that the mod operation is applied to each element of the matrix. The elements $u_{k,l}$ of the matrix $U_{h,v}$ represent the amount of change in a corresponding pixel. Now the watermark $f_{i,j}$ can be defined as follows.

$$f_{i,j} = \begin{cases} 1, u_{i,j} \neq 0 \text{ and } d_{i,j} \neq 0 \\ 0, \text{otherwise} \end{cases} \tag{7}$$

The watermark is embedded into a location where the value of $f_{i,j}$ is one, that is, where the magnitude of the watermark is none zero.

Let $ID = \{id_0, id_1, \cdots, id_{l-1}\}$ be a text that indicates the copyright information. In the following Eqn.(8), C_{ID} is obtained by repeating ID a number of times. The repetition is necessary in order to reduce/eliminate errors during extraction.

$$C_{ID} = \underbrace{ID\|ID\|\cdots\|ID},\qquad\qquad (8)$$
$$\phantom{C_{ID} = } L$$

Here L is the number of locations where the value of $f_{i,j}$ is one. We note that clearly repetition can be replaced with a more efficient error correction code.

C_{ID} is then randomized with $m_{i,j}$, an M-sequence with a maximum period. This ensures that the condition 3) discussed above is satisfied. We denote the randomized sequences by $s_{i,j}$ which are defined more precisely as follows:

$$s_{i,j} = \begin{cases} id_y \oplus m_{i,j}, & \text{if } f_{i,j} = 1 \\ m_{i,j}, & \text{otherwise} \end{cases} \qquad (9)$$

where id_y is an element in C_{ID} .

Finally a watermark pattern $w_{i,j}$ is constructed via Eqn.(10) and it is embedded into the original image. The watermarked image I_W is obtained by the addition of the pattern $w_{i,j}$ to a corresponding pixel in the original image.

$$w_{i,j} = \begin{cases} (2 \times s_{i,j} - 1) \cdot u_{i,j}, & \text{if } f_{i,j} = 1 \\ 0, & \text{otherwise} \end{cases} \qquad (10)$$

To extract the watermark, we subtract the original image from the watermarked image. Then, a value $\hat{s}_{i,j}$ is derived from the signs of $u_{i,j}$ and $\hat{w}_{i,j}$ of Table 1.

Table 1. The extraction of $\hat{s}_{i,j}$

	$\hat{w}_{i,j} > 0$	$\hat{w}_{i,j} < 0$
$u_{i,j} > 0$	1	0
$u_{i,j} < 0$	0	1

To reconstruct the copyright information ID , $\hat{id}_y$ is extracted using $m_{i,j}$ and $\hat{s}_{i,j}$ as in Eqn.(11).

$$\hat{id}_y = \hat{s}_{i,j} \oplus m_{i,j}, \text{if } f_{i,j} = 1, \qquad (11)$$

If $\hat{s}_{i,j}$ is equal to $s_{i,j}$ used in embedding, the extracted $\hat{id}_y$ should be the same as the original id_y. As a result, we are able to reconstruct the copyright information ID correctly.

4   Simulation Results

We simulated the proposed method to evaluate the robustness of the following block data which serves as a simple example.

$$\begin{pmatrix} 20 & 20 & 53 & 79 & 80 & 56 & 21 & 20 \\ 20 & 82 & 110 & 110 & 110 & 110 & 86 & 22 \\ 53 & 110 & 110 & 110 & 110 & 110 & 110 & 59 \\ 79 & 110 & 110 & 110 & 110 & 110 & 110 & 85 \\ 80 & 110 & 110 & 110 & 110 & 110 & 110 & 86 \\ 56 & 110 & 110 & 110 & 110 & 110 & 110 & 62 \\ 21 & 86 & 110 & 110 & 110 & 110 & 91 & 23 \\ 20 & 22 & 59 & 85 & 86 & 62 & 23 & 20 \end{pmatrix}$$

For the above data, the visual components can be easily calculated. These components are shown below, and indicate the edges components in a block.

$$\begin{pmatrix} 2 & 2 & 3 & 3 & 3 & 3 & 2 & 2 \\ 2 & 3 & 4 & 3 & 2 & 3 & 3 & 2 \\ 3 & 4 & 3 & 2 & 2 & 2 & 3 & 3 \\ 3 & 3 & 2 & 0 & 0 & 0 & 2 & 3 \\ 3 & 3 & 2 & 0 & 0 & 0 & 2 & 2 \\ 3 & 3 & 2 & 0 & 0 & 0 & 2 & 2 \\ 2 & 3 & 3 & 2 & 2 & 2 & 2 & 2 \\ 2 & 2 & 3 & 3 & 2 & 3 & 2 & 2 \end{pmatrix}$$

Consider a single bit "1" as copyright information ID to be embedded into a block. The differences and the watermark patterns for a JPEG quality level $q_1 = 25$ are shown in Figure 2.

The length L indicates the number of locations where the watermark is embedded. For the above data, its value is 59 out of 64. That is, 59 data among 64 are used for embedding the watermark. When these watermarked data are compressed with a quality level q_2 that is greater than q_1, we are able to reconstruct the copyright information if the signs of the extracted watermark are the same as those of the original watermark.

In another experiment we use $q_2 = 80$. The results are depicted in Figure 3. To reconstruct the embedded single bit "1" copyright information, the embedded random sequences $s_{i,j}$ are derived from Table 1 by using the signs of the extracted watermark and pattern calculated by Eqn.(6). The id_y can be calculated from $s_{i,j}$ using Eqn.(11). The embedded ID can be extracted by counting the number of 1's and 0's in id_y followed by a majority vote. In the simulation, the numbers of 1's and 0's are 31 and 28, respectively. This gives us "1" as the value of the embedded ID. Note that more efficient error correcting codes, rather than simple repetition, can also used.

In this example, the proposed method is able to reconstruct the copyright information with respect to $q_2 = 80, 75$, but not to $q_2 = 65 \sim 20$, as the size of the block is too small. In general, the larger the size of an image, the higher the chance of extraction. This motivates us to apply the proposed method to the standard image Lena(256×256 , 8bits/pixel).

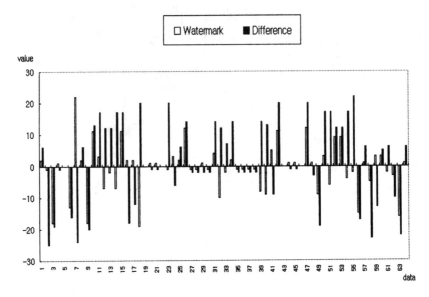

Fig. 2. The difference and watermark pattern for $q_1 = 25$

The watermarked image is shown to Figure 4. We used $q_1 = 25$ as the quality control parameter and the length of copyright information is 112 bits (14 digit ASCII characters). The watermark is embedded into 60,870 pixels among 65,536. Table 2 indicate the numbers of 1's and 0's, for the first 10 bits "0100110000" of the 112 bits.

Table 2. The numbers of 0's and 1's from the extracted bits used to construct *ID*

quality level	bit order (the num. of 0 : the num. of 1)				
q_2	1	2	3	4	5
80	357:187	201:343	354:190	355:189	197:347
75	329:215	219:325	343:201	339:205	213:331
60	315:229	226:318	310:234	314:230	208:336

quality level	bit order (the num. of 0 : the num. of 1)				
q_2	6	7	8	9	10
80	189:355	356:188	345:199	350:194	344:200
75	206:338	336:208	343:201	325:219	327:217
60	229:315	312:232	325:219	315:229	302:242

From Table 2, one can reconstruct correctly all the bits. Table 3 clearly indicates that it is possible to reconstruct the copyright information when the watermarked image is compressed with a quality level of q_2 greater than q_1 .

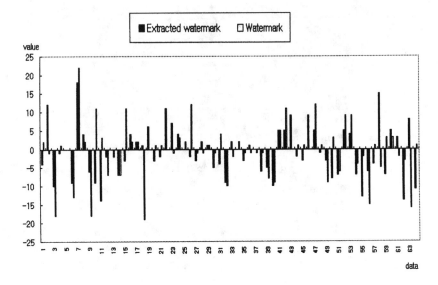

Fig. 3. Extracting a watermark for q_2

Table 3. The reconstruction of ID for $q_1 = 25$

quality level q_2	80	75	60	35	30	25	20
reconstruction	○	○	○	○	○	○	×

5   Conclusion

The proposed method utilizes the quality level of JPEG as a parameter in water-marking to provide robustness against JPEG compression. It calculates visual components using relationships with neighboring pixels. Locations to embed the watermark are derived from these visual components. To minimize extraction errors caused by compression, copyright information is repeated and converted into random sequences by XOR with M-sequences. As the result, the owner can extract the watermark even when the watermarked image is compressed, as long as the quality level is not smaller than the quality level used as a parameter. The copyright information can be extracted from the embedded random sequences at a later stage.

To close this paper, we remark that in this work we have considered only the quality level of the JPEG compression algorithm as a parameter. One may use different parameters related to image processing, and these parameters might provide equal or even stronger robustness against various attacks based on image processing.

Fig. 4. The watermarked image for $q_1 = 25$

Acknowledgment : *This work was supported by GRANT No. KOSEF 981-0928-152-2 from the Korea Science and Engineering Foundation.*

References

1. M.D.Swanson, M.Kobayashi and A.H.Tewfik, *Multimedia Data-Embedding and Watermarking Technologies*, In Proc. of IEEE, Vol.86, No.6, pp.1064-1087, 1998
2. M.M.Yeung et al, *Digital Watermarking*, Communications of the ACM, Vol.41, No.7, pp.31-77, 1998
3. C. T. Hsu and J. L. Wu, *Hidden Signature in Images*, In Proc. of IEEE International Conference on Image Processing, pp.223-226, 1996
4. G.Langelaar, J.van der Lubbe and J.Biemond, *Copy Protection for Multimedia Based on Labeling Techniques.*,
   http://www-it.et.tudelft.nl/pda/smash/public/benelux_cr.html
5. M.Kutter, F.Jordan and F.Bossen, *Digital Signature of Color Images Using Amplitude Modulation*, In Proc. SPIE-EI97, pp.518-526, 1997
6. K.Matsui and K.Tanaka, *Video-steganography: How to Embed a Signature in a Picture*, In Proc. IMA Intellectual Property, Vol.1, No.1, pp.187-206,1994
7. I.Cox, J.Kilian, T.Leighton and T.Shamoon, *Secure Spread Spectrum Watermarking for Multimedia.*, Tech.Rep., 95-10, NEC Research Institute, 1995
8. C.I.Podilchuk and W.Zeng, *Image-Adaptive Watermarking Using Visual Models*, IEEE Journal on Selected Areas In Communications, Vol.16, No.4, pp.525-539,1998
9. J.Dittmann, M. Stanenau and R. Steinmetz, *Robust MPEG Video Watermarking Technologies*, ACM Multimedia'98, 1998
10. A.B.Watson, *DCT Quantization Matrices Visually Optimized for Individual Images*, Proc. of SPIE, 1992

11. E. Koch and J. Zhao, *Towards Robust and Hidden Image Copyright Labeling*, Proc. of IEEE Workshop on Nonlinear Signal and Image Processing, 1995
12. J.Fridrich, *Methods for Data Hiding*, http://ssie.binghamton.edu/jirif/

Fingerprints for Copyright Software Protection

Josef Pieprzyk

Center for Computer Security Research
School of Information Technology and Computer Science
University of Wollongong
Wollongong, NSW 2522, Australia
josef@uow.edu.au

Abstract. The work studies the problem of copyright protection of software using cryptographic fingerprinting. The identity of software must be unique for each copy. There are two classes of identities: one is based on equivalent variants of the program and the other applies behaviour of the software as its identity. For these two types of identity, we introduce two different fingerprint schemes. The two schemes use digital signatures and can be easily combined and extended to be resilient against partial fingerprint destruction.

1 Introduction

The ease in which electronic documents can be duplicated and distributed over communication networks (Internet), makes the copyright piracy as easy as execution of a copy operation. Exchange and sharing of data enhances productivity and efficiency of our work. Finding a balance between sharing and protection of electronic data seems to be difficult in the Internet environment. Once a copy of a copyrighted work (music, picture, animation, video, software, books) is available in electronic form, the work can be copied arbitrary number of times. Moreover, a copy is indistinguishable from the original.

Protection of electronic documents against illegal copying has become of outmost importance to dynamically growing companies selling their products such as audio and video CDs, software and other copyrighted products. Electronic commerce (E-commerce) supports not only the commercial transactions but also enables the products to be shipped to the buyer via Internet. There has been many papers dealing with protection of digital audio and video documents (for instance see [4, 7]). Software protection is studied in [1, 3, 6].

2 Motivation

Software protection relates closely to watermarking of electronic documents (such as picture, video, animation, sound, or a composition of those used in a typical web page). Watermarking heavily depends on the ability of some electronic documents to tolerate a small "watermarking" noise. On the other hand,

M. Mambo, Y. Zheng (Eds.): ISW'99, LNCS 1729, pp. 178–190, 1999.
© Springer-Verlag Berlin Heidelberg 1999

the software does not tolerate any change and its integrity is crucial for correct functioning. A single bit change may render the software useless. Therefore the techniques used for watermarking of documents are not, in general, applicable to software protection.

The need for software protection may arise from a composition of legal and commercial factors. Certainly, the creator (producer, author) of the software may need to make sure that her software package can be told apart from other similar packages accessible on the market in the case of legal proceedings resulting from a fault in a software whose authorship may be questionable. An efficient and secure watermarking of software may also be useful for verification of the source of the software in case when it is covered by a warranty or perhaps, the producer has guaranteed the buyer a free update of the software.

On the other hand, there is a booming black market of cheap software packages which are simply illegal copies of renowned software companies. Although, sometimes it is impossible to eliminate the distribution of fake software (mainly, because not all countries are signatories of the copyright convention), producers may try to help buyers to enable them to verify the source of the software.

One could expect that some people will try to copy their legally acquired software to share it with friends or perhaps to resell it illegally. Clearly, if the buyer (owner) of the software makes copies, it is impossible to distinguish them from the original held by the buyer.

3   Software Copyright Protection

The following are three broad categories of software protection methods:

1. watermarking,
2. fingerprinting,
3. software identification protocols.

Watermarking techniques are widely applicable for multimedia data protection. A watermarked document is intentionally distorted by adding a noise whose characteristic can be uniquely attributed to the producer of the document. Note that to remove the watermark, in most cases, it suffices to superimpose an additional noise – this typically removes the watermark in the expense of lowering the quality of information [3, 4].

A fingerprint is a variant of digital signature. Software is treated as a document on which the producer generates their signature using a cryptographically secure signature scheme. A nice feature is that everybody can verify the copyright ownership of the fingerprinted software. Also the security features may be proved [2].

The previous two methods treat a piece of software as a passive entity. They may have some advantages but certainly they do not allow to exploit the fact that software is active entity. The active character of software can be used to incorporate fingerprints into the state of the software or the piece of software can identify itself (using a zero-knowledge protocols [8], for instance).

In general software copyright protection is an identification problem. To make an identification method work, one must first ensure that objects can be uniquely identified. Note that electronic copy of any software gives two or more objects which are indistinguishable. The only solution seems to be to allow single (and identifiable) objects (software copies) to exist. If we assume that each copy of the same software package is different, then a sheer fact of existence of two identical copies indicates a copyright infringement.

Identification of software can be performed by

1. the copyright holder who designed the software and knows all details about the software (its structure and hidden information),
2. the buyer of the software who knows almost nothing about internal structure of the software but may know some public information about the alleged producer (copyright holder).

Clearly, the design of software protection seems to be easier when the verifier is the copyright holder. This makes sense if the software is distributed via authorised dealers where the buyers do not need to worry about illegal copies as the vendor takes care of software protection.

In the advent of E-commerce where software can be bought via Internet, even if the buyer uses strong authentication during the transaction, they still need to verify whether the software shipped via Internet is the correct one (i.e. with agreed specification and made by specific company). This is especially important if the bought software must be compatible to the already installed operating system and made by the specific producer.

4   Software Watermarking

Software watermarking is concerned with different methods of hiding information which identifies the producer or creator (holder of copyright). Typically, to demonstrate the presence of a watermark, the producer invokes the program for a (secret) parameter. The program returns the watermark which could be for instance, *Copyright by John Smith Ltd.* Note that watermark can be recovered only the the copyright owner.

There are two main watermark types depending on how watermarks are embedded into software (program). They are as follows ([3]):

1. Static watermarks – an identifying string (watermark) is stored into variables declared in the program.
2. Dynamic watermarks – a watermark is built up in real time by the program during its execution.

Static watermarks are easy to identify and remove as they are stored explicitly within the program. Dynamic watermarks are not readily identifiable as their presence is not evident by inspecting the source code. Unfortunately, a good quality debugging tools can be used to inspect dynamic structures constructed by the program and detect the watermark. If a dynamic watermark is detected, the attacker may try to remove a part of the code which is responsible for its construction. This is always possible especially if the watermark code is not functionally tied up with the rest of the program.

Consider a dynamic watermark scheme described in [3]. The copyright holder selects two large primes p, q and computes their product $n = p \times q$. The integer n is encoded in a graph structure. The graph with encoded n constitutes a watermark. When the copyright holder wishes to prove that a piece of software contains her watermark, she links an additional program called recogniser and run the software for a secret input. The recogniser reads the encoded n from the graph and returns it as the watermark. As she is only person who knows the factors, she can claim she has proven herself to be the copyright holder. The solution suffers from the following weaknesses:

- The watermark can be extracted and incorporated into an illegal copy of some other software.
- The watermark can be modified after disabling the so-called tamperproofing.
- The copyright holder can prove the ownership once only by devolution of the two primes. This weakness could be readily fixed by allowing the copyright holder to use a zero-knowledge protocol [8] for proving $n = p \times q$ without revealing the factors.

5 Identity of Software

We argue that copyright protection is normally implemented as an identification protocol in which the copyright holder can prove her ownership to the software. A piece of software can be identifiable only if any single copy has its own identity embedded into it. Again, we assume that the existence of identical copies of the same software indicates forgery. An identity can be incorporated into a piece of software using the following approaches.

- Code variants – because of an inherit redundancy of any computer program (no matter whether it is a source or object code), it is possible to create different copies of the same software. Note that all copies exhibit the same functionality.
- Behaviour of software – different copies are created using different and unique "behaviour" of software. For instance, each program can have a unique collection of static and dynamic data structures. Exception handling can be unique for any copy.
- Tamperproof module – the identity (secret) plus a part of software code is stored on a tamperproof smart card which at the beginning of interaction, verifies integrity of the remainder of the code.
- Remote module – software is split into two parts: one resides on the users machine and the other is accessible as a remote procedure. The remote procedure is controlled by the copyright holder.

The above approaches for identity embedding are independent in the sense that identity can be imposed by any combination of them.

Code variants can be easily generated for any software as long as the creator of software prepares a list of variants for each instruction. Note that identity can be also easily altered by swapping these instructions whose order does not influence the functionality. As variants are performing the same operations, this does not require any expertise or analysis of the software.

Identity via customising behaviour of software seems to offer better security (especially against non-experts) in the expense of increased complexity for the software design and implementation. This approach can be as trivial as assigning a unique collection of variables used in the program or as complex as design of unique collection of modules (procedures, functions) with their unique data structures and codes. The overall functionality stays the same.

The underlying assumption about tamperproof components is that their contents cannot be seen or modified without destroying or damaging the components. To make illegal copy of software, possession of a duplicate of tamperproof module must be available. This approach makes sense if the user who wants to make multiple copies of a copyrighted software, is not able to bypass the module. This is to say that the tamperproof module must contain crucial information or functions without them the software renders useless or the construction of the bypass module is approximately as expensive as the design of the whole software from scratch.

Most machines are already connected to Internet. The copyright holder may sell software in two parts: one is given to the buyer (user) and the other remains with the owner. Needless to say that the remote piece of software is accessible to the user (for instance via a remote procedure call facility). Again, the software should be designed in such a way that the piece in user's possession must be useless without the second held by the owner. This is attractive option especially if the software is rented for some period of time or the user is charged for usage of the software. Also if a typical user does not have enough resources to run efficiently some calculations but the copyright owner has a powerful machine, then it is reasonable to leave the resource-consuming parts with the owner and allow users to use them remotely.

5.1   Group Software Identity

Assume that there is a software package which includes several programs. It is possible to embed identity for each piece independently of other pieces. The verification of identity must then be done independently for each piece. Similarly, assume that we have a very long (single) program. The identity could be verified by checking the integrity of the whole program. One would ask whether the verification process can be reduced. The answer is positive and the solution involves secret sharing. Let us describe the idea. Consider a software which consists of n pieces and the copyright holder wants that any t identities (out of n) are enough to identify the whole software (the group). To implement such group identity, it is enough to design n identities using (t, n) threshold scheme.

Note that the verification of the software can be done by considering a randomly selected subset of t program pieces. The improvement in efficiency trades off with the probability of a mistake in verification.

5.2   Software Identity Resilient against Partial Destruction

It is not unreasonable to assume that identity can be lost after a small code modification. It is possible to apply secret sharing so if some parts of the code are tampered with, the true identity can still be recovered. As before secret sharing comes to the rescue. If it is expected that only a part of the program code will be tampered with, then the whole code is divided into n pieces. Each piece is assigned an identity according to shares of (t, n) threshold scheme. Any t identities allow to determine the program identity. In case of illegal modification of the program, the program identity can be recovered if there are t program pieces with undisturbed identities. Clearly, it is desirable to be able to identify modified pieces of program (by for instance using message code authenticators).

6   Security Requirements for Software Protection

There are four major players: the copyright holder (producer), a user (buyer or owner), an attacker (or enemy) and the software (program). A buyer during purchasing transaction typically needs to verify whether the software delivered by a vendor is correct, i.e. issued by the alleged copyright holder (software manufacturer) and is of the required type and specification, etc. The verification must be publicly available as potentially every body can be a buyer. The probability of false acceptance must be negligible. The user would not like to get a fake software. From the user point of view the probability of false rejection is not of a primary concern. Clearly, the copyright holder or a vendor are interested in minimising the probability of false rejection.

The copyright holder needs to identify her software in the following circumstances:

- a user requests a service (software upgrade) for the program. The copyright holder may use the publicly available verification,
- an attacker distributes illegal copies of her software. Clearly, the publicly available identification will not work as the attacker (knowing the identification scheme) can destroy software identity. In this case, it can be desirable for the copyright holder to incorporate a hidden identification scheme which is kept secret until the need arises (the proof of the ownership in the court of law). Note that this identity verification can be done once only as after the proof, the secrecy is compromised.

Obviously, a software protection scheme must address concerns of users and copyright holders. In other words, users would like to have a public identification scheme (fingerprint) while copyright holders would prefer secret embedding of fingerprint. Additionally, the copyright holder would like to claim the software ownership even after some illegal modification.

7   Nested Fingerprinting

The first solution is based on public key cryptography and applies nested signature schemes to enable both the user and the copyright holder to prove identity of software. First we have to show how software identity is defined.

7.1   Identity and Mutable Instructions

The copyright holder must make sure that any copy of the same software has its unique identity. Consider an object code. Most machine instructions can be written in many equivalent forms. For instance, given two operands X and Y. To sum them, it is possible to use the following two equivalent forms

        LOAD X; ADD Y;

or

        LOAD Y; ADD X;

which translates: load the operand into accumulator and later add the operand to the accumulator. The equivalent instructions have been already widely used by writers of computer viruses to make viruses mutate after each infection.

Similarly, most instructions in a specific high-level programming language (such as Pascal, C/C++, Java) can be written in many equivalent forms. For instance, a substitution

        x:=y;

can be rewritten as

        x:=(y+2)-2;

or perhaps as a sequence of two substitutions

        x:=y+1; x:=x-1;

In C, the substitution x=y could be easily presented equivalently as x=(y++)-1; or x=(y--)+1;. Similarly, conditional expressions could be readily converted into their equivalents by modifying the condition.

For a given piece of software, there are two classes of instructions: immutable (when the instruction has no variants), mutable (when the instruction has two or more variants). Depending on circumstances, immutable parts may be lines of code such as declarations of variables, functions, procedures, etc. which, for obvious reasons, should be left the same in all copies of the software.

Definition 1. *Given a mutable instruction $e_i \in S$ of a code S. Then the instruction e_i can be represented equivalently by*

$$e_{i,1} \equiv e_{i,2} \equiv \ldots \equiv e_{i,\ell}$$

Denote

$$\mathcal{E}_i = \{e_{i,j}; j = 1, \ldots, \ell\}$$

as the set of instructions equivalent to e_i. The set $\mathcal{E}_i$ is also called the variant collection.

There are two possibilities as far as variants are concerned. In the first one, the collection of variants is independent from the position of the specific instruction within the code. In result, all copies of a software preserve the sequence of instructions. The second possibility enables the creator of a software to manipulate versions in the context of the code. She may generate versions by changing the order of instructions if this does not influence the correctness of the software. For the sake of simplicity we assume that all mutable instructions have variants independent from their positions in the code (the first case).

Definition 2. *Given a variant collection $\mathcal{E}_i = \{e_{i,j}; j = 1, \ldots, \ell\}$. Assume that each variant $e_{i,j}$ is assigned its unique variant number (or variant label) j or*

$$V(e_{i,j}) = j.$$

If the cardinality of $\mathcal{E}_i$ is ℓ, then the set $\mathcal{Z}_\ell = \{1, \ldots, \ell\}$ contains all labels associated with $\mathcal{E}_i$. Note that $V : \mathcal{E} \to \mathcal{Z}_\ell$.

Definition 3. *Given a sequence of mutable instructions $S = (e_1, \ldots, e_n)$ with $e_i \in \mathcal{E}_i$; $i = 1, \ldots, n$. Then the sequence*

$$id_S = (V(e_1), \ldots, V(e_n))$$

gives the identity sequence of the code S.

Definition 4. *Two copies S', S'' of the same piece of software (code) are considered to be different if*

$$id_{S'} \neq id_{S''}.$$

Definition 5. *The collection of all different copies of a single software S is denoted by $\mathcal{S}$.*

Given a piece of software $S = (e_1, \ldots, e_n)$ with mutable instructions $(e_1, \ldots, e_n)$; $i = 1, \ldots, n$. Then the number of different copies of S which can be produced is

$$|\mathcal{S}| = \ell_1 \times \ldots \times \ell_n$$

where $e_i \in \mathcal{E}_i$ and $|\mathcal{E}_i| = \ell_i$; $i = 1, \ldots, n$.

7.2   The Scheme

Assume that the producer holds its unique secret key k_p. There is also a trusted authority (TA) who keeps the matching public key K_p in a form of a publicly accessible certificate (the public key is signed by TA using its secret key so every body can verify its integrity). There is also a cryptographically strong collision-free hash function $H : \Sigma^* \to \Sigma^m$ with public description. The function takes a message of arbitrary length $M \in \Sigma^*$ and yields an m-bit digest $d = H(M) \in \Sigma^m$. The producer applies a cryptographic signature $SG_p : \Sigma^* \to \Sigma^r$ by $SG_p(M) = D_{k_p}(H(M)) \in \Sigma^r$.

Consider a software $S = (\mathcal{E}_1, \ldots, \mathcal{E}_n)$ and divide it into two parts: the body $S_b = (\mathcal{E}_1, \ldots, \mathcal{E}_r)$ and the tail $S_t = (\mathcal{E}_{r+1}, \ldots, \mathcal{E}_n)$.

Definition 6. *Given a software $S = (S_b, S_t) \in \mathcal{S}$ and its producer with the pair of keys (K_p, k_p). Then the fingerprint of S is*

$$FP_S = (id_{S_b}, id_{S_t})$$

such that $id_{S_b} = D_{k_p}(H(S_M, id_{S_t}, id_p, \text{edition date, vendor id}))$ where id_{S_t} is randomly selected and is unique for the given copy of software. S_M is a master copy of the software which is used to produce all fingerprints. The identity of the producer is id_p.

The master copy can be seen as a variant of the software. This variant does not have any fingerprints and typically includes original instructions. Note that fingerprints defined above are simply digital signature.

The verification of the fingerprint follows verification steps for signatures. We assume that the collection of variants for all mutable instructions is publicly known in the form of the variant table $\mathcal{V}$. To verify the originality of a copy S, the verifier first recovers id_{S_t} from the copy and the publicly known variant table $\mathcal{V}$ and recomputes

$$\tilde{H}(S_M, id_{S_t}, id_p, \text{edition date, vendor id}).$$

Next, the verifier identifies the signature $id_{S_b} = D_{k_p}(H(S_M, id_{S_t}, id_p, \text{edition date, vendor id}))$ from the copy and knowing the public key of the producer recovers the hash value from the signature so

$$H(S_M, id_{S_t}, id_p, \text{edition date, vendor id}).$$

Finally, the copy is considered original if the two hash values are identical or $\tilde{H} = H$.

Now we are ready to describe the nested fingerprinting scheme. There are three basic operations defined: embedding fingerprints, identity verification by user and by the copyright holder.

Embedding Fingerprints

Assumptions are as above, i.e. there is already public-key infrastructure so public keys of copyright holders are available from the TA in the form of certificates. There is also a collision-free hash function H with public description. A cryptographically strong pseudorandom bit generator PBG is used to make internal fingerprints invisible for users (and attackers). The embedding is done by the copyright holder and runs according to the following steps:

1. Preparations – given a software $S \in \mathcal{S}$ normally in the form of its master copy S_M and a security parameter which defines the size of the block used to divide S. The software S is divided into $2un + 1$ blocks so $S = (S_1, \ldots, S_{2un+1})$. The parameter u can be selected depending on the length of the software and security requirements.
2. generating signature – compute $h = H(S_M, id_p, \text{edition date, vendor id})$ and the corresponding signature $SG_p = D_{k_p}(h)$.

3. Constructing secret sharing – create u pairs of (t, n) threshold schemes (the size of $GF(q)$ is defined by the security requirements). The first secret sharing in each pair shares the value h and the second one – the digital signature SG_p. In result, there are $2un$ shares (blocks). Let the sequence of shares be $\alpha = (a_1, \ldots, a_{2un})$.

4. Masking shares – the copyright holder chooses a secret β and generates a long enough string using the pseudorandom bit generator so

$$\gamma = (\gamma_1, \ldots, \gamma_{2un}) = \alpha \oplus PBG(\beta)$$

5. Incorporating internal fingerprints – for each block of software S_i; $i = 1, \ldots, 2un$, its variant S_i^* is computed such that its identity $id_{S_i^*} = \gamma_i$.

6. Incorporating external fingerprints – A hash value is computed for the sequence of all variants or $h^* = H(S_1^*, \ldots, S_{2un}^*, id_p$, edition date, vendor id). The variant of the last block is selected in such a way that its identity is $id_{S_{2un+1}^*} = SG_p(h^*)$.

Note that each block of software S_i must be long enough to incorporate $2n$ shares each of the length $\log_2 q$ bits.

Identity Verification by User

Any user can verify the fingerprint by checking whether the identity of the last block matches the signature $G_p(h^*)$ i.e. she checks whether the hash value h^* obtained for the first $2nu$ blocks is the same as the hash value recovered from the signature embedded into S_{2un+1}^*.

Identity Verification by Copyright Holder

The copyright holder can prove the identity of the software by first computing the string $PBG(\beta)$, taking away the masking and recovering $2n$ collections of shares $\alpha = \gamma \oplus PBG(\beta)$, and verifying signatures in each collection of $2n$ blocks. To pass the check, it is enough to have t undisturbed blocks in each collection of n blocks.

Security obviously relates to the security of signature scheme, hashing and pseudorandom bit generator. The nested fingerprinting provides en external fingerprint publicly verifiable. Internal fingerprints are masked and in fact the string γ is polynomially indistinguishable from a truly random for a cryptographically strong PBG. An attacker is not even aware of the existence of internal fingerprints.

There is an interesting question, whether the copyright holder may falsely claim ownership to some software by trying to play with secret β to obtain a desired α. The problem can be stated as follows. Given $\gamma = \alpha \oplus PBG(\beta)$ and a random γ'. Is it possible to find β' such that

$$\alpha = \gamma \oplus PBG(\beta) = \gamma' \oplus PBG(\beta') \ ?$$

The answer is no for any random γ' much longer than the seed β'. To see this it is enough to count the number of possible seeds versus the number of possible random γ'.

7.3   Implementation Considerations

Most computer software is sold and distributed as object codes. Fingerprints can be embedded by the compiler. Note that optimising compilers try to remove instruction redundancy. Fingerprinting compilers add instruction redundancy and as such can be perceived as the "negation" of optimising compilers. The proposed scheme can also be used for source codes but compiler will certainly remove fingerprints (especially when optimised ones are used). In this case the ownership can be proved by showing the fingerprinted source code.

8   Behavioural Fingerprinting

The previous solution uses object or source codes to incorporate fingerprinting. A random selection of variants for each instruction will certainly strip off all the fingerprints. An alternative solution would be to use behaviour of the software as identity. In other words, the master copy of a software gives a generic behaviour which is defined by the application. Each copy must behave differently although the generic behaviour must be retained.

8.1   Dynamic Identity

A software S is considered during the run time so it can be modelled as a Deterministic Turing (DT) machine with its transition function $\delta : Q - \{q_N, q_Y\} \times \Gamma \to Q \times \Gamma \times \{-1, +1\}$ where Q is a finite set of states with two halting states $\{q_N, q_Y\}$ and Γ is a finite set of tape symbols (for details see [5]).

Definition 7. *The identity of software S modelled by its DT machine (DTM_S), is a sequence of states generated from the initial state q_T using an input string x_T. Denote $id_S = (q_1, \ldots, q_n)$.*

To extract identity from a software, it is enough to set the initial state to q_T and run the executable for the input x_T. The sequence of states resumed throughout n steps defines the identity. To differentiate identities for copies of the same software, the producer must incorporate redundancy into the states. The difficulty of removing fingerprints very much depends on how the redundancy is incorporated. Specially designed states which are traversed only in the identity testing process seems to be easily designed and removed. Better solution is to add redundant states which are also used in the "normal" execution.

8.2   The Scheme

We again use cryptographic techniques to create fingerprints. There is a TA which distributes public keys of copyright holders on demand in the form of certificates. So the public keys of copyright holders are known. The producer (copyright holder) uses a cryptographic signature SG_p. A collision free hash function H is publicly available.

Embedding Fingerprint

1. Create a unique variant S of the software $\mathcal{S}$ by adding redundant states with its identity $id_S = (q_1, \ldots, q_n)$.
2. Compute $h = H(S_M, id_S, id_p, \text{additional info})$ and calculate $SG_p = D_{k_p}(h)$. Incorporate the signature SG_p into S by modifying it so $id_{S'} = (id_S, SG_p)$.

The verification process can be made public if both the starting verification state q_T and the input x_T are public. The verification can be restricted to the producer only if the verification parameters are secret. There is nothing to prevent the copyright holder to incorporate multiple fingerprints or a fingerprint which is resilient against partial destruction (using secret sharing).

Instead of security proof, we would like to emphasise that the above fingerprinting can be removed from the software but it will cost the attacker some effort to identify all fingerprints and rewrite the software so it cannot be claimed by the copyright owner. The amount of work really depends on the expertise of the attacker and the way the fingerprints are embedded into the software.

9   Conclusions

Nested text fingerprints can be easily combined with dynamic fingerprints forcing a potential attacker to rewrite the whole software. To remove dynamic fingerprints, the attacker must change the behaviour of the software. This means he must reset most of the variables and remove those which are redundant (as they are likely to be used to hold fingerprints). This redesign process may turn to be more expensive than designing a new software from scratch.

Acknowledgement

We would like to thank the anonymous referees for their constructive criticism and improvements to this paper.

References

[1] D. Aucsmith. Tamper resistant software. In R. Anderson, editor, *Information Hiding, First International Workshop, Cambridge, 1996*, pages 317 – 334. Springer-Verlag, 1996. Lecture Notes in Computer Science No. 1174.

[2] D. Boneh and J. Shaw. Collusion-secure fingerprinting for digital data. In D. Coppersmith, editor, *Advances in Cryptology - CRYPTO'95*, pages 452–465. Springer-Verlag, 1995. Lecture Notes in Computer Science No. 963.

[3] C. Collberg and C. Thomborson. On the limits of software watermarking. Technical Report 164, Department of Computer Science, the University of Auckland, Auckland, New Zealand, 1998.

[4] I. Cox, J. Kilian, T. Leighton, and T. Shamoon. A secure, robust watermark for multimedia. In R. Anderson, editor, *Information Hiding, First International Workshop, Cambridge, 1996*, pages 183 – 206. Springer-Verlag, 1996. Lecture Notes in Computer Science No. 1174.

[5] M. Garey and D. S. Johnson. *Computers and Intractability: A Guide to the Theory of NP-Completeness.* Freeman, 1979.

[6] A. Herzberg and S. Pinter. Public protection of software. *ACM Transactions on Computer Systems*, 5(4):371–393, 1987.

[7] J. Smith and B. Comiskey. Modulation and information hiding in images. In R. Anderson, editor, *Information Hinding, First International Workshop, Cambridge, 1996*, pages 207 – 226. Springer-Verlag, 1996. Lecture Notes in Computer Science No. 1174.

[8] D.R. Stinson. *Cryptography: Theory and Practice.* CRC Press, 1995.

A Secrecy Scheme for MPEG Video Data Using the Joint of Compression and Encryption

Sang Uk Shin, Kyeong Seop Sim, and Kyung Hyune Rhee

Department of Computer Science, PuKyong National University,
599-1, Daeyeon-dong, Nam-gu, Pusan, 608-737, Korea
shinsu@woongbi.pknu.ac.kr, seaside@unicorn.pknu.ac.kr,
khrhee@dolphin.pknu.ac.kr
http://unicorn.pknu.ac.kr/~{soshin, seaside, khrhee}

Abstract. We propose a fast encryption method for MPEG video which encrypts the sign bits of DC coefficients and uses a random scan instead of zig-zag one. The proposed algorithm is a method which combines an encryption with MPEG video compression. Furthermore, it has short processing times for the encryption and provides sufficient security from the viewpoint of visibility and signal to noise ratio values. Thus the proposed method is suitable for the protection of the moving pictures with minimizing the overhead of MPEG video codec system. Also, we consider an authentication method appropriate for stream data such as MPEG video.

1 Introduction

Multimedia data security is an important issue in multimedia commerce. For example, in video on demand(VOD), it is desirable that only those who have paid for the services can view their videos. In video conferencing, it must be possible to view the video only between the specific groups of video conferencing.

One method of secure distributed multimedia applications is to encrypt multimedia data, but it requires large computations. Generally, the features of multimedia data is that their data size is very large and it needs to be processed in real-time. Processing vast amounts of data puts a great burden on video codec.

Multimedia data, specially MPEG video has characteristics of large data size and relatively low information value compared with text data. Due to such characteristics, to encrypt the entire data is an inefficient method. Hence it needs to process the encryption/decryption with small computation.

In this paper, we propose an efficient algorithm for encrypting the MPEG video. This algorithm encrypts the sign bit of DC coefficients and use a random scan instead of a zig-zag one. The proposed algorithm is a joint method which combines an encryption with MPEG video compression, and it provides a sufficient security with small processing time.

The remainder of this paper is organized as follows. In section 2, we shortly describe MPEG video compression and in section 3, we discuss the related works. In section 4, we propose a new MPEG video encryption algorithm and analyze

M. Mambo, Y. Zheng (Eds.): ISW'99, LNCS 1729, pp. 191–201, 1999.

the algorithm in section 5. And we consider the existing authentication method of MPEG video in section 6. Finally, we have conclusions in section 7.

2 The Principle of MPEG

2.1 The Structure of MPEG-1 Video

MPEG-1 is composed of a sequence of groups of pictures(GOPs) which is randomly accessible and is a quantity of about 0.5 seconds. Usually, each GOP is consist of 15 pictures and each picture is one among I, P and B picture.

I pictures are intraframe coded without any reference to other pictures. P pictures are forward-predictively coded using a previous I or P picture. And B pictures are bidirectionally interpolated from both the previous and following I and/or P pictures. Fig. 1 shows the structure of GOP.

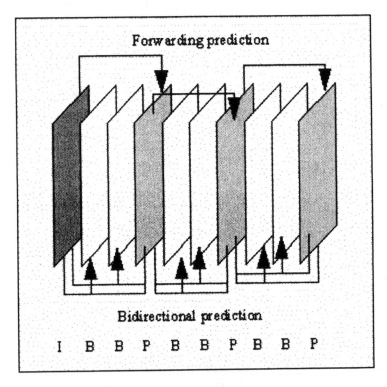

Fig. 1. The structure of GOP

Each picture is divided into macroblocks. A macroblock is a 16×16 pixel array. Regardless of the type of pictures, each macroblock is further subsampled into four 8×8 luminance blocks(Y blocks), two 8×8 chrominance blocks(a Cr block and a Cb block). Each block is composed of a DC coefficient and 63 AC coefficients.

2.2   The MPEG-1 Video Compression

The three important techniques of MPEG video coding are the transform coding, the motion compensation and the entropy coding[1]. As shown in Fig. 2, each of the 8×8 Y, Cb and Cr blocks is fed to a DCT(Discrete Cosine Transformation), quantization and Huffman entropy coding[4].

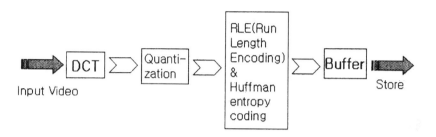

Fig. 2. The MPEG-1 video compression

DCT concentrates most of the energy in the upper-left corner of the 8×8 block[4]. After the quantization process, many DCT coefficients become zero. By the zig-zag scan order, an 8×8 pixel block is linearized to a 1×64 pixel vector. And the run length encoding(RLE) turns a 1×64 pixel vector into a sequence of (skip, value) pairs. Then the Huffman entropy coding compresses the (skip, value) pair sequence. Specially, the DC coefficient of I pictures is separately coded from the AC coefficients. In this paper, we process the encryption of MPEG video data with the minimized overhead by encrypting the sign bit of DC coefficients and changing the zig-zag scan order randomly.

3   Related Works

Previous cryptography studies have focused on the text data and many encryption algorithms for the text data have developed. However, they can not be directly applied to multimedia data because of the following reasons: First, it has large data size. Thus it involves heavy computations. Second, it requires a real-time processing.

A simple method to secure MPEG video[2] is to encrypt the entire data with DES(Data Encryption Standard)[3]. However it is too slow to meet the real-time requirement of MPEG video playback. A method to solve this problem is to selectively encrypt the portion of the MPEG video[6][7] : encrypting MPEG video header, portions of I, P and B pictures or both. Another method is to replace the zig-zag scan order with a random scan order[1]. In a 8×8 block of each picture, it uses a random permutation list to replace the zig-zag scan order which maps the DCT coefficients to a 1×64 vector.

In this paper, we use both the selective method and the random permutation list. The proposed algorithm provides a sufficient security with a minimum of the encryption of bits and the processing time. The proposed scheme generates the random permutation list in real time and applies a different permutation list to each GOP in order to increase the security. And the sign bits of DC coefficients are encrypted. This method is secure against existing known plaintext attacks and ciphertext attacks.

4   The Combination of the MPEG Compression with the Selective Encryption

In our scheme, the purpose of the encryption process is to encrypt the extremely small portions of the entire video stream in order to reduce the computations in the encryption/decryption process, and to make it secure against known attacks.

The proposed algorithm is consist of two parts. The first part is to use the random permutation list instead of zig-zag scan order. Due to this process, it cannot get the exact values in IDCT(Inverse DCT) of the decompression process and make the picture blurry. The other part is to encrypt the sign bits of DC coefficients. It causes the exact values of the block to be not able to be computed in the decompression process.

We use the RC4 algorithm as a random permutation function to generate the random permutation list. This process can be executed during the compression/ decompression of the video stream without the delay in real-time applications.

In the encryption process of sign bits(S) of DC coefficients of I pictures, video stream S is applied to the encryption function E_k which is inversible using secret key k, and the ciphertext C is generated ($C = E_k(S)$). A user with the key k can get the original pictures by using its inverse $E_k^{-1}(C)$. In this process only 6 bits per a macroblock is extracted. The extracted bits are encrypted by using the encryption algorithm such as DES, RC4[3] or RC5[3]. This can reduce the overhead during the encoding/decoding process.

5   The Result and the Analysis of Simulation

5.1   Simulation Environment

We simulated the proposed algorithm on Pentium MMX(200MHz). We modified the avi2mpg1 for Windows 95/NT of MPEG Simulation Group[9] and used the Windows Media Player.

To generate the random permutation, we modified RC4[3][5]. And 128-bit RC4 stream cipher was used for the encryption. To increase the security, a block cipher such as RC5 can be used.

5.2   Analysis of the Result

The proposed scheme encrypts the small portions of the entire stream and uses the random permutation list. In Fig. 3, a circle-dot is the number of encrypted bits when the entire stream is encrypted, and a square-dot is a number of encrypted bits when the proposed method is used. As the result of the experiment, about 1.2% of the entire stream was encrypted. Comparing with the encryption of all or parts of sign bits of DC and AC coefficients, the proposed scheme showed better encryption effect which the output images were not distinguishable visually.

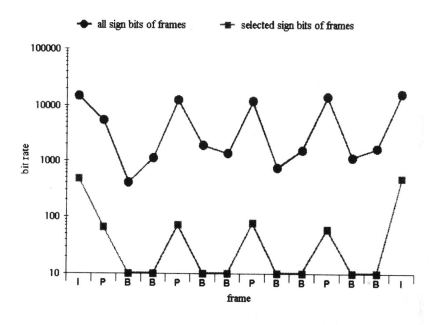

Fig. 3. The number of encrypted bits per a frame

Table 1 compares the file size of the zig-zag scan order with that of the random permutation list. When using the random permutation list, the file size increases about 0.3% on an average. During playing the MPEG video, it leads a negligibly small overhead.

Fig. 4 shows original images which are arranged in a counter clockwised direction. Fig. 5 shows the images applied random permutation lists. In the case of Fig. 5, images are distorted on the whole but one can observe, even though it is not obvious, some information on the original images. Thus the method using only a random scan has a drawback for the encryption of images.

Fig. 6 shows the images that the sign bits of DC coefficients are encrypted by random sequences. In these images, one can not obtain obvious images but one can partially see the blurry original image. Thus this scheme may also involve some weakness. Fig. 7 is the images that the proposed method which combines an encryption of sign bits with the random scan is applied. One can not distinguish the original image at all.

Table 1. The file size of the zig-zag scan order and the random permutation list

	the result of the zig-zag scan	the result of the random permutation list
1	1,800,397	1,798,672
2	1,271,772	1,282,646
3	1,832,113	1,847,673
4	1,692,900	1,693,412
5	1,691,678	1,692,370
6	1,081,348	1,081,695
7	1,415,653	1,415,509
8	1,831,660	1,847,193
9	1,554,137	1,567,483
10	1,691,695	1,707,281
11	1,573,772	1,574,122
12	1,360,938	1,361,117
13	2,008,546	2,007,421
14	3,210,136	3,210,891

Fig. 4. The original images

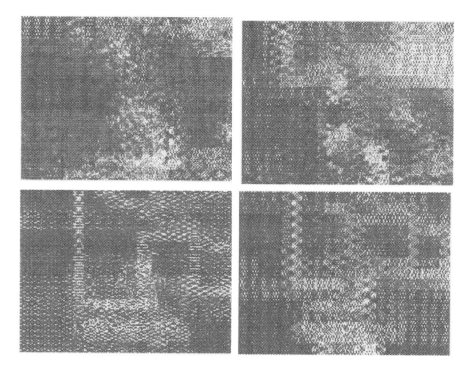

Fig. 5. The images applied random permutation lists

As a more objective measure, we compared the SNR(Signal to Noise Ratio) values of the random-scanning method, the method of encrypting sign bits of DC coefficients and the proposed method which combines an encryption of sign bits with the random scan, respectively. For computer simulations, we tried a test for 28 images. As the results, the average SNR value of the proposed method for 28 images is 0.868 while the random scanning method is 4.344 and the method of encrypting sign bits of DC coefficients is 1.264, respectively. In Fig. 8, we can see that the proposed method is more effective one than other two methods by comparing SNR values.

As another experiment, when encrypting the DC value and a few AC values, one can not distinguish the original image at all, but the number of the encrypted bits should be increased. An encryption only for AC values can not induce a satisfiable result. Therefore, the proposed scheme is more efficient than the method which encrypts DC values or AC values.

A small change of DC value has an large effect on encryption. In this paper, we encrypted only the sign bit of DC value and scanned AC values by the random permutation list. Because it has very small computations, this scheme is very useful for real-time applications in which much data need to be encrypted. Also, in the proposed algorithm, the encryption process is combined into the compression process. Thus the proposed method may be used as a new efficient method combining an encryption into the compression.

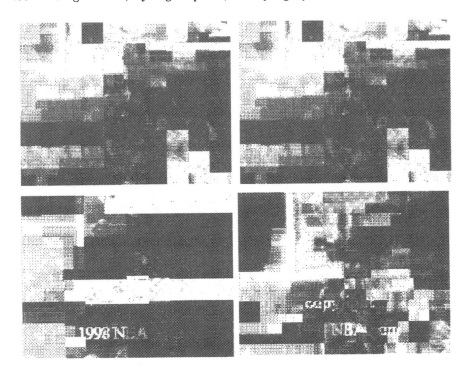

Fig. 6. The images encrypted only the sign bits of DC coefficients

6   MPEG Video Authentication

In this section, we consider the authentication method of MPEG video in [10]. The problem of signing digital stream is substantially different from the problem of signing regular message. We assume that the sender knows the entire stream in advance and MPEG data is logically divided into blocks. The receiver first receives the signature on the n-byte hash of the first block(if a dedicated hash function is used, n is usually 20-byte). After verification of the signature, the receiver knows what the hash value of the first block should be and then starts receiving the full stream and computing its hash block by block. If the hash value does not match, it rejects the block and stops playing the stream. Other blocks are authenticated as the following : the first block contains the n-byte hash value of the second block, the second block contains the n-byte hash value of the third block and so on ... Thus only the hash values for every subsequent block are checked after checking the signature of the first block. The detail algorithm is as the following:

Let (S, V) be a regular signature scheme. The sender has a pair of secret-public key (SK, PK) of such signature scheme. And let H be a collision-resistant hash function. The GOP data of MPEG video stream is

$$B = B_1, B_2, \ldots, B_k$$

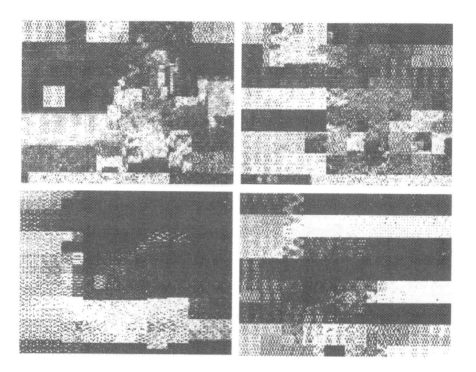

Fig. 7. The images using the proposed methods

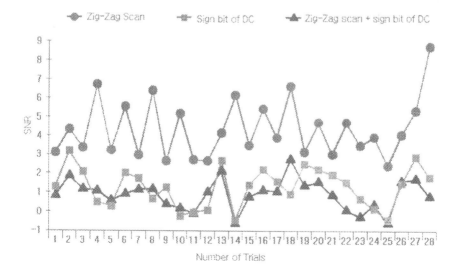

Fig. 8. The comparison of SNR

and the resulting signed stream is

$$B' = B'_0, B'_1, B'_2, \ldots, B'_k$$

first, B_k is signed, e.g., the processing is done backwards. Each block has the hash value of the next block.

$$B'_k = < B_k, 00\ldots0 >$$

$$B'_i = < B_i, H(B'_{i+1}) > \qquad for \quad i = 1, \ldots, k-1$$
$$B'_0 = < H(B'_1, k), S(SK, H(B'_1, k)) >$$

On the sender side, computing the signature and embedding the hashes requires a single backwords pass on the stream, hence the stream has to be fully known in advance. Also the first signed block B'_0 contains the length of the stream(k). For the verification of the signature, the receiver checks the following on receiving $B'_0 = < B, A_0 >$:

$$V(PK, A_0, B) = 1$$

and extracts the length k in the block. Then on receiving $B'_i = < B_i, A_i >$ (for $1 \le i \le k$) the receiver accepts B_i if

$$H(B'_i) = A_{i-1}$$

Thus the receiver has to compute a single public-key operation at the beginning and then only one computation of a hash value per block. No big table is needed in memory.

In the case of MPEG video constructed by above method, one can authenticate the entire MPEG video by considering practical methods for embedding authentication information as the followings: as a first method, MPEG video has a USER-DATA section where the authentication data can be placed. Secondly, the MPEG system layer allows for an elementary data stream to be multiplexed synchronously with the packetized video stream. Such elementary stream can carry the authentication information. A third scheme uses digital watermarking to embed information in video stream at the cost of slight quality degradation.

As a further research, we will study a method for providing both encryption and authentication of MPEG video by embedding the hash value into the area where it has the minimum quality degradation by using a digital watermarking.

7   Conclusions

In this paper we proposed a fast encryption algorithm of multimedia data, specially, MPEG-1 video, which can be applied to real-time applications. The proposed method is a new efficient one with very small computations that achieves the encryption/decryption in real time. By encrypting the sign bit of DC value of I picture which has most of information in MPEG data, we achieved a sufficient security in visual point of view as well as SNR values with very small overhead.

Moreover, a random permutation of the zig-zag scan order of AC values per a GOP increases the security and makes the images invisible. The proposed scheme may be used for the secrecy of MPEG video such as VOD, video conferecing applications and a limited TV system. For a further research, we introduced a suitable method for the authentication of stream such as MPEG video data, and we considered its efficient implementation method.

References

1. L. Tang: Methods for Encrypting and Decrypting MPEG Video Data Efficiently, In Proceedings of the ACM Multimedia96, Boston, MA., (November, 1996) 219-229
2. L. Qiao, K. Nahrstedt: Comparison of MPEG Encryption Algorithms, International Journal on Computer&Graphics, Special Issue: "Data Security in Image Communication and Network", Vol.22 No. 3, published bimonthly by Permagon Publisher (January, 1998)
3. B. Schneier,: "Applied Cryptography", John Wiley & Sons (1996)
4. J. David Irwin: "Emerging Multimedia Computer Communication Technologies", Prentice Hall PTR (1998)
5. S. Mister: Cryptanalysis of RC4-like Ciphers, Master's thesis, Queen's University, Kingston, Ontario (1998)
6. C. Shi, B. Bhargava: A Fast MPEG Video Encryption Algorithm, In Proceedings of the 6th ACM International Multimedia Conference, Bristol, UK (September, 1998)
7. C. Shi, B. Bhargava: An Approach to Encrypt MPEG Video in Real-time to be appeared in ICMCS99 (June, 1999)
8. http://www.kjmbc.co.kr/beta/digital/comp_3.html
9. MPEG Software Simulation Group (MSSG), http://www.mpeg.org/MPEG/MSSG/
10. R. Gennaro, P. Rohatgi, how to sign digital stream, Proceedings of CRYPTO'97, Santa Babara, CA, USA, (Aug, 1997)

On Anonymous Electronic Cash and Crime

Tomas Sander and Amnon Ta–Shma

International Computer Science Institute
1947 Center Street, Berkeley, CA 94704, USA
{sander,amnon}@icsi.berkeley.edu

Abstract. Anonymous electronic cash systems can be vulnerable to criminal abuse. Two approaches were suggested for solving this problem. In an escrowed cash system users are given anonymity which can be lifted by trustees. In an amount–limited system each user enjoys unconditional anonymity but only for payments up to a limited amount during a given time-frame. In this paper we discuss the two approaches.

1 Introduction

The Problem. In 1982 David Chaum [3] showed how to build an anonymous electronic cash system by devising blind signature schemes. Chaum's scheme is provably anonymous: even an all powerful agent that collaborates with the bank and any coalition of the users can not link payments to withdrawals, i.e. payers enjoy unconditional anonymity. In 1992 van Solms and Naccache [13] discovered a serious attack on Chaum's payment system. Blackmailers could commit a "perfect" blackmailing crime by using anonymous communication channels and anonymous ecash. Following that, further concerns were raised, e.g., it was argued that the ability to move money around anonymously at the speed of light may facilitate money laundering activities and tax evasion.

Escrowed Cash. These concerns stimulated a whole line of research of, so called, escrowed cash systems In escrowed systems payment transactions look anonymous from the outside (to users, merchants, banks), while Trustees are able to revoke the anonymity of each individual payment transaction.

Does Escrowed Cash Solve the Problem? We describe several weaknesses of escrowed cash in Section 2. We demonstrate that using some simple tricks criminals may still be able to hide their suspicious activities in an escrowed system in a way that is hard to detect. We further argue that escrowed cash is not a natural solution to some of the major attacks on electronic cash systems (blackmailing and bank robbery) that, in our opinion, are caused *not* by the anonymity feature but rather stem from the fact that most anonymous cash systems are implemented using signature based schemes. Finally, from the honest user's point of view escrowed cash might be undesirable, as no user can ever be sure of his privacy.

An Alternative Approach. We then discuss in Section 3 a different approach that was suggested by us in [11, 10]. We conclude in Section 4.

M. Mambo, Y. Zheng (Eds.): ISW'99, LNCS 1729, pp. 202–206, 1999.
© Springer-Verlag Berlin Heidelberg 1999

2   Weaknesses in Escrowed Cash Systems

Revoking Anonymity Only Under Suspicion. In the U.S. the banking system has to report any cash transaction above $10,000 to law enforcement agencies. These high value payment activities are then monitored for "suspicious" transactions. According to FinCEN monitoring high volume transactions is an important and essential tool for *triggering suspicion* and fighting crime. On the other hand, in escrowed cash systems anonymity should be revoked only under suspicion and thus the order of suspicion and monitoring is reversed. It seems that the current proposed schemes for escrowed cash are in this respect much weaker then techniques currently used in fighting financial crimes today.

Issuer Attacks. In many offline or escrowed cash systems a payee can deposit received coins only back into the bank, unless he has some additional knowledge of a secret known to the payer (e.g. double spending related). We give an example of an issuer related attack in which a bank cooperates with a drug dealer L. L has received a large amount of electronic money (e.g. stemming from drug sales) in a currency issued from this bank that he does not want to deposit into his bank account to avoid being questioned about its origin by agencies monitoring account activities. Instead of depositing the payment transcripts into his account he exchanges the payment records secretly against fresh, unreported coins of the electronic cash issuer. Thus L has received spendable electronic money in return for his payment transcripts. From the bank's point of view the balance is preserved. In (non-escrowed) offline systems the transaction is almost riskless for the bank [1]. Such transactions become more risky and difficult if the bank has to keep a publicly verifiable and complete whitelist of withdrawal transcripts [2].

Blackmailing and Signature-Based Systems. In the blackmailing attack [13], the attacker forces the bank to issue valid coins via anonymous communication channels that are indistinguishable from valid coins, and thus can not be later recognized by the bank as stemming from a crime. In the bank robbery attack [7], the secret key the bank uses for signing coins is stolen, and the attacker issues valid unreported money. While the attacks use the anonymity feature of the system, the anonymity feature by itself is not enough for the attacks to work. This is obvious, e.g., with the bank robbery attack, where the attack assumes the bank has a secret key for signing coins. We believe that these attacks show the vulnerability of blind-signature based schemes, rather than of anonymous schemes in general.

Some papers [7, 4, 9, 6] suggested to use the escrow features for detecting such an attack when it goes on, or for fixing it later on, but not for preventing it efficiently altogether. However, even detection is not easy in escrowed cash systems, and [6] claim to be the first(!) to provide detection at a very early stage

---

[1]   Even double spending risks are small, as the bank can always move some of the payment transcripts received from L into its official database, and then the double spender can be detected as usual.

[2]   See [1, 5] for reports on laundering activities that involved the (sometimes unintentional) cooperation of financial institutions.

that a bank key has been compromised. Fixing the situation usually requires the system to move to an on-line mode and many times requires rebuilding of the whole system. Many systems (but not e.g. [6]) also have the additional unpleasant feature that fixing the system requires lifting the anonymity of all users and coins. Instead of seeing this as a motivation for escrowed cash one may conclude that blind–signature based techniques might not be well suited for anonymous electronic cash.

The Colombian Black Market Exchange. Here we describe a weakness that is due to the fact that money is *transferable*. The attack already exists in completely non-anonymous systems. It seems also to exist in escrowed cash systems.

The so called "Colombian Black Market Peso Exchange" is described in the FATF 1997-1998 Report on Money Laundering Typologies [2] as an "example of a widely used money laundering technique". Its goal is that drug traffickers in the U.S. want to transfer profits from drug sales to Colombia. Simplified it can be described as follows [3]. Colombians residing in or visiting the U.S. open several personal check accounts at U.S. banks. The customers sign blank checks for these accounts and give them to the drug cartels. Then drug money in low amounts is transferred to these accounts. Later a dollar amount is entered into the check but the name of the payee is left blank. These checks are later sold in Colombia to Colombian business men who need dollars to conduct business in the U.S. at a discounted rate for Pesos. The Colombian businessman fill in the payees name and can use the checks for payment. It is important to notice that the attack works even in a completely non-anonymous environment, like checks and U.S. bank accounts. The crux of the attack is that the checks can be made *transferable*.

3   An Alternative Approach

A different approach to limit potential abuses of anonymous electronic cash systems was suggested by the authors in [11, 10]. We observe:

- Financial transactions that use the banking system as an intermediary are easier to monitor. This monitoring has been described as an important tool against money laundering. Banks are closely and regularly monitored to ensure safety and soundness of the financial system.
- Law enforcement agencies are interested in large value transactions. Small value transactions are usually not monitored.

We define the following (technical) requirements for electronic cash systems:

1. **Amount-Limited Anonymity.** [11] Each user can anonymously spend only a fixed amount, say $ 10,000 a year, and transactions exceeding this amount do not enjoy the anonymity protection.
2. **Non-transferability.** [11] Only the user who withdraws a coin can make a payment with it. The payee can only deposit the coin back into the bank.

---

[3] cf also [12] for a detailed presentation of this attack.

3. **Auditability.** [10] There should be a one-to-one correspondence between withdrawal records and valid coins.

Non-transferability. The example of the "Colombian Black Market Peso Exchange" attack makes direct use of checks with blank payee that can be obtained by one person and used for payment by another, i.e. of the transferability feature of checks (resp. electronic cash). It seems that a non-transferable system is not vulnerable to this attack. [4]

Amount–Limited Anonymity.   Financial crimes typically involve large amounts of money. Thus, it is a natural approach to limit the amount of anonymous electronic cash that users (and criminals) can obtain. This can be achieved by limiting the amount of anonymous electronic cash each user can obtain in a *non-transferable* system. We observe that the non-transferability requirement is essential; in a transferable system criminal organizations may obtain electronic cash in large amounts by buying it from other users.

Auditability. To protect against issuer related attacks it is desirable to have an electronic cash system in which the *money supply* can be closely monitored, and issuers can not secretely issue electronic coins. The auditable system in [10] is not based on "blind signatures" but on the primitive of "blind auditable membership proofs". In this system, the security of the system does not rely on the secrecy of secret keys of the bank, but on the bank's ability to maintain the integrity of a *public* database. As a result, it is no longer vulnerable to the bank robbery attack. Furthermore, there is no way to force the bank to issue coins in the system [10], that can not be invalidated, a property that is important for defending against blackmailing of the bank.

4   Conclusions

Starting with [13] several potential abuses of completely anonymous electronic cash systems were found. Escrowed cash was introduced to solve this problem. We believe that escrowed cash systems, at least partially, do not live up to this goal. We have pointed out several weaknesses in escrowed cash systems.

- Escrowed cash systems are hard to secure against the blackmailing and the bank robbery attack.
- Escrowed cash systems do not per se address issuer related attacks.
- Black markets and exchanges may effectively circumvent tracing.
- Escrowed cash does not effectively *find* suspicious activities and law enforcement may be harder than today.

One way to solve these problems is, in our opinion, by using auditability to solve issuer related attacks, blackmailing and bank robbery, and non-transferability and amount-limitedness to address money-laundering and tax evasions. Furthermore, our solution guarantees users provable, unconditional

---

[4] Another case where non-transferability can be useful is for the implementation of the so called Geographic Targeting Order. See [1, 8].

anonymity. Yet, users are limited in the amount of anonymous electronic coins they can use per time frame and they can not transfer their anonymous money.

Chaum showed that modern cryptography makes unconditionally anonymous electronic cash possible. On the other hand, completely anonymous electronic cash systems might be abused for unlawful activities. Thus, one should first determine whether he/she finds these abuses threatening enough to require a solution. It seems that if one wants to address these possible abuses, one needs to make some kind of a compromise. This raises two questions. First, do we prevent these abuses by making the compromise? and second, what is the price we pay for the compromise? Comparing the escrowed cash approach and our approach with regard to the second question, one could ask oneself what is better: "unlimited amounts of money with revokable anonymity" or "limited amounts of money with unconditional anonymity".

Some aspects of the solution we suggest in [10] are only theoretically efficient but not yet practical. It would be interesting to see a practical and privacy friendly implementation of these ideas.

References

[1] FATF-VII report on money laundering typologies, August 1996. http://www.treas.gov/fincen/pubs.html.

[2] FATF-IX report on money laundering typologies, Febuary 1998. http://www.ustreas.gov/fincen/typo97en.html.

[3] David Chaum. Blind signatures for untraceable payments. In *Crypto 82*, pages 199–203, 1982.

[4] E. Fujisaki and T. Okamoto. Practical escrow cash system. *Lecture Notes in Computer Science*, 1189, 1997.

[5] General Accounting Office (GAO). Private banking: Raul Salinas, Citibank, and alleged money laundering. GAO/OSI-99-1, December 1998. http://www.gao.gov/monthly.list/dec98/dec9811.htm.

[6] M. Jakobsson and J. Muller. Improved magic ink signatures using hints. In *Financial Cryptography*, 1999.

[7] M. Jakobsson and M. Yung. Revokable and versatile electronic mony. In *3rd ACM Conference on Computer and Communications Security*, pages 76–87, 1996.

[8] R.C. Molander, D.A. Mussington, and P. Wilson. Cyberpayments and money laundering. RAND, 1998. http://www.rand.org/publications/MR/MR965/MR965.pdf/.

[9] H. Peterson and G. Poupard. Efficient scalable fair cash with off-line extortion prevention. *Lecture Notes in Computer Science*, 1334, 1997.

[10] T. Sander and A. Ta-Shma. Auditable, anonymous electronic cash. In *Crypto*, 1999.

[11] T. Sander and A. Ta-Shma. A new approach for anonymity control in electronic cash systems. In *Financial Cryptography*, 1999.

[12] Bonni G. Tischler. The Colombian black market Peso exchange. Testimony before the Senate Caucus on International Narcotics Control, June 1999. http://jya.com/bmpe062199.htm.

[13] S. von Solms and D. Naccache. On blind signatures and perfect crimes. *Computers and Security*, 11(6):581–583, October 1992.

On the Difficulty of Key Recovery Systems

Seungjoo Kim[1,*], Insoo Lee[1,*], Masahiro Mambo[2], and Sungjun Park[1,*]

[1] KISA (Korea Information Security Agency),
5th FL., Dong-A Tower, 1321-6, Seocho-Dong, Seocho-Gu, Seoul 137-070, Korea
{skim, insoo, chaos}@kisa.or.kr
http://dosan.skku.ac.kr/~sjkim/
http://dosan.skku.ac.kr/~sjpark/
[2] Education Center for Information Processing,
Tohoku University, Kawauchi Aoba Sendai, 980-8576 Japan
mambo@ecip.tohoku.ac.jp

Abstract. Key escrow cryptography has been becoming popular recently. A key escrow system bridges the gap between users' privacy and social need for protection against criminal behavior. However, there are some disadvantages and controversies regarding the system. In this paper we review and analyze the weaknesses of several recent protocols. The protocols are examined with respect to their claimed issues.

1 Introduction

Key escrow cryptography has gained much attention recently. The history of key escrow started in April 1993, with the proposal by the U.S. government of the Escrow Encryption Standard [37], EES, also known as the *CLIPPER* scheme. But the original idea came earlier, at least as early as in 1992 when Micali introduced the concept of *fair public-key cryptosystems* [34]. The most controversial issues about key escrow or key recovery are that

- It is difficult to ensure that the key escrow system will function correctly and not be modified by the user to bypass or corrupt the escrow process (see 2, 3 of §2.1 and 1, 2 of §2.2).
- Furthermore, key escrow may restrict his/her privacy and give controlling power to society, which may, in certain circumstances, abuse it (see 1 of §2.1, 2, 3 of §2.2, and §2.3).

In this paper, we review requirements and concerns of several technical proposals, and point out their designs do not present the required property. This paper is organized as follows. Section 2 describes categories of attacks and their possible countermeasures. Section 3, 4, 5 and 6 review and analyze the protocols of [33,16,28], and [22]. In addition, section 7 and 8 present some new subliminal attacks. Finally, section 9 contains concluding remarks.

---

* Supported by MIC (Ministry of Information and Communication in Korea) under project 98-260-07.

M. Mambo, Y. Zheng (Eds.): ISW'99, LNCS 1729, pp. 207–224, 1999.
© Springer-Verlag Berlin Heidelberg 1999

2   Related Work

Attacks to the key recovery system can be divided into three main categories : attacks against the user security component, attacks against the recovery agent component, attacks against the data recovery component, i.e. law enforcement agent. [17] These logical components are highly interrelated, and the design choices for one affect the others. The possible attacks against each of these components is described in the followings.

2.1   Attacks when Encrypting a Message in Transit

Generally, the sender of a message performs three functions with respect to key recovery [58] : (1) Obtain key recovery encryption device/information, (2) Encrypt the message with the key recovery information, (3) Prove key recovery compliance to the receiver or third party. Thus the possible attacks against the user security component can be decomposed into the above three categories :

1. **Obtain the key recovery encryption device/information :**

   - *Kleptographic attack* : The art of stealing user information secretly and subliminally in a cryptographic black-box device, i.e. in the U.S. government Clipper and in particular in Capstone escrow technology. [53,54,55]

2. **Encrypt the data with the key recovery information so that future data recovery is possible :**

   - *Double(over-/under-) encryption attack* : pre- or post- encryption of the traffic with another cipher. [10,11]
     Impossibility of countermeasure : If one can derive a predicate distinguishing encrypted message from unencrypted message, the double encryption attack can be detected. However, such a predicate does not exist : One can easily encode encrypted messages so that they look like natural language, e.g. "zero zero one zero ... ". [42]
   - *Pfitzmann-Waidner's attack on binding ElGamal* : The attack is similar to the general double encryption attack. The variation is that an attacker can use the key recovery scheme itself for the inner encryption [42] :

   $$m =< info||E_{S'}(m')||E_{P_{Bob}}(S') >,$$

   $$c =< E_S(m)||E_{P_{Bob}}(S)||E_{P_{Trusted\_Recovery\_Party}}(S)||bind > .$$

   - *Chaffing and winnowing* : When sending a message, add fake(chaff) packets with bogus MACs. The chaff packets have the correct overall format and reasonable message contents, but have MACs that are not valid. To obtain the correct message, the receiver merely discards all of the chaff packets, and retains the wheat packets having valid MACs. [43]

− *Abuses of probabilistic encryption schemes* : Criminals who want to send
  a secret message to a receiver can abuse any escrow encryption systems
  based on probabilistic cryptosystems by simply generating each ciphertext block whose parity, e.g. number of 1's, is exclusive-or of one bit of
  the intended message and one bit of Vernam cipher's key shared between
  the criminal and the receiver. [52]

3. **Prove key recovery compliance to the receiver or third party :**
   − *Universal verification of recoverability problem* : If there is no way to
     check whether a LEAF, Law Enforcement Access Field, actually con-
     tains the right session key then, by sending noise instead of a LEAF, the
     abuse is possible. This can be prevented by the verifier's validation or re-
     calculation of LEAF prior to decryption. We call this receiver-verification
     of recoverability.
     (a) *Receiver-verification of recoverability* : The receiver can verify that
         authorities can recover message from the ciphertext. Thus, it allows
         law enforcement access as long as a criminal interoperates with legal
         peer.
     However, abuse by a colluding of a sender and receiver is still easily
     possible. So we need a way to publicly verify and assure that the session
     key or the private-key is escrowed properly, i.e. public-verification of
     recoverability.
     (b) *Public-verification of recoverability* : Everybody, not only the partic-
         ipants, can verify the correctness of LEAF.
   − *Brute-force LEAF search attack* : Since the checksum contained in the
     128 bit LEAF of Clipper is only 16 bits in length, any randomly generated
     128 bit string will have a $\frac{1}{2^{16}}$ (non-negligible) chance of appearing valid.
     [10,11]
     Countermeasures : To frustrate this attack, (1) EES devices can be mod-
     ified to limit the number of incorrect LEAFs they will accept. How-
     ever, this approach increases the vulnerability of the system to denial-
     of-service attacks. (2) Alternatively, a more robust solution is to increase
     the size of LEAF checksum to 32 or 64 bits, making exhaustive search
     infeasible.

   The followings are restricted to an attacker communicating with other
   rogue systems.

   − *LEAF feedback attack* : Upon negotiation of a session key, the receiving
     side of a rogue application can simply generate a new IV/LEAF and feed
     it back to itself, where the sender have never sent the IV/LEAF at all.
     This still leaves the problem of IV synchronization. However, by using
     dummy messages compensating for error propagation, we can solve this
     problem. [10,11]
     Countermeasure : In the LEAF feedback attack, two entities bypass the
     system by dropping the LEAF from communication. This means they

didn't follow the compliance certification, so, by monitoring the communication of dishonest attackers, the two entities can be caught.

– *LEAF obscuring attack* : Encryption of the LEAF. Since a key exchange protocol produces "extra" shared secret bits such as Diffie-Hellman, the shared bits but 80 bits for Skipjack session key can be used as a Vernam cipher against the LEAF. [10,11]

– *Self-squeezing attack* : This attack strengthens Blaze's LEAF feedback attack in that, using the self-squeezing attack, their communication will look like they follow the compliance certification, i.e., their ciphertext has a LEAF. But, after the recovery process, all law enforcement will notice is that the messages sent are randomly looking while they are not. (the entities may claim they were merely testing !) To launch a self-squeezing attack, the rogue entities generate two keys K and K' in the key exchange phase, then they each generate a $LEAF'$ using the key K'. A sender uses the actual key K to generate encryption and true $LEAF$, and before transmission he just replaces the generated true $LEAF$ with the $LEAF'$. Before decrypting, a receiver replaces the sent $LEAF'$ with his own true $LEAF$. The rest is the same as the case of LEAF feedback attack. [21]

Countermeasure : To protect against self-squeezing, one can design the scheme to have the session key be a function of the message and the message originator and receiver ID. Compliance can be verified under some mechanism like a tamperproof device of EES.

2.2   Enrollment Attacks against the Key Recovery Agent

The enrollment process at the KRA, Key Recovery Agent, breaks down into five steps [58] : (1) The KRA obtains the user's key recovery secret, (2) The KRA binds the key recovery secret to the user's ID, (3) The KRA stores the secret, (4) The KRA retrieves the secret and provides it to an authorized agent, (5) The KRA destroys the secret after it is no longer needed. Depending on the systems some of these steps do not exist.

1. **The KRA obtains the user's key recovery secret :**
   – *Shadow public-key attack* : A subliminal attack on a given public-key. The attacker generates two key pairs (P, S) and (P', S'), where (P, S) is a proper (public-key, private-key) pair, (P', S') is a shadow key pair, and $P' = f(P)$ where f is an easily computed and publicly known function. In order for someone to send a secret message to an attacker, the sender computes $P' = f(P)$ and then encrypts the message using P'. [25]

   Countermeasure : To defend against this attack, the KRA and the user interactively select the private-key.

   – *PKI-compatible problem* : In most threshold-escrow systems except one in [56] and [57], every user must split his private-key and send shares to the authorities. The authorities must then interact, i.e. stay in-line, to ensure that the key is escrowed. The system thus has more communication overhead than a typical public-key system. A PKI-compatible system should be designed such that (1) no interaction is needed and (2) even if the KRA needs more computation, the user's amount of computation is similar as the case of PKI.

2. **The KRA binds and stores the key recovery secret to the user's ID :**
   - *Key cloning attack* : This allows an entity who is an intentional future suspect to copycat a key of another user. Then when this entity misbehaves, a court may issue a wiretapping order, which will enable law enforcement to listen to the honest user as well (due to cloning). [21]
   Countermeasure : The KRA is a possible entity from which the suspect obtains the key. By employing the technique of *protected function sharing* [1] at the KRA we can design a technique where the user's key is not required to be shared to agents.
   - *Advance warning attack* : KRA can give advance warning to the person whose line is being tapped or to be tapped.
   Countermeasure : Thus KRA, who is cooperating with the investigator to eavesdrop a user's communication, should not be able to know whom the investigator is intercepting. We can realize such a system by applying a blind decoding protocol. [46]

The followings occur when a sender himself binds the key recovery secret to the identity information.

   - *Squeezing attack* : An attack to use other people's LEAFs. Criminals can squeeze an honest person's LEAF in order to force escrow agents to open the honest person's escrow key. So, users who have no part in some illegal actions whether knowing it or not may have their keys opened and messages read. [21]
   Countermeasure : To protect against the squeezing attack, one can design to have the session key be a function of the sender and receiver ID, i.e. ID-based key distribution. Compliance can be verified under some mechanism like the tamperproof device.
   - *Spoofing attack* : In environments like the internet with both fair cryptosystems and Clipper, two collaborating parties can spoof two legitimate users and can frame them, by putting their unescrowed session key into the message header of two legitimate parties. [21]
   Countermeasure : An authentication, e.g. a digital signature protection, is needed with key recovery system to assure the source liability to its message and for prevention of spoofing. However, generally dealing with the worst case of spoofing seems hard.

---

[1] See §2.4 for detail.

3. **The KRA retrieves the secret and provides it to an authorized agent :**
    - *Large scale wiretapping problem* : Even if the government is trustworthy today it may be replaced by an un-trustworthy government in the future. This government may, for reasons of its own, suddenly start un-escrowing the secret keys of all users and embark on mass wiretapping. [48,44,5,6] *Countermeasure* : To prevent the government from a large scale decryption, a user's private-key c is divided into two parts x and a such that $c = x + a$, where a is a number of short length and x is the escrowed key. When the LEA, Law Enforcement Agent, intends to monitor the user, only more than certain number KRAs can collaborate to recover x. But to get c, he must carry out brute search for a. [2]
    - *Multiple domain problem* : It should be compatible with the different laws and regulations of participating countries concerning interception, use, sale and export. [23,14,36]
    - *Perfect-forward-secrecy problem* : A system with forward secrecy is one in which compromising the keys for decrypting one communication does not reduce the security of other communications. Key recovery destroys the property of forward secrecy in many key recovery systems. [1]

2.3   Attacks against Authorized Access

The key recovery system is designed to provide authorized access to encrypted messages. The process breaks down into three steps : (1) Authorized agent proves warranted access. For example, an LEA provides a court order to the KRA, (2) The KRA retrieves the key recovery secret and provides it to the LEA, (3) The LEA decrypts the cipher.

- *Time and target limitation problem* : This is consistent with "minimization rule" policy. [19]
    1. *Time-boundness* : It must be possible for the courts to enforce the time-limits of a warrant, by supplying a key that will be effective only for a given period of time (most likely some set of days). [29,13]
    2. *Target flexibility* : It must be possible for the courts to permit either [29]
        (a) *Node surveillance* : All communications involving a particular target A can be decrypted,
        (b) *Edge surveillance* : Only communications between A and B can be decrypted.
- *Monitoring-of-messages-to- and not-from-criminals problem* : Fair cryptosystem requires the opening of the keys of individuals receiving messages from criminal rather than opening the keys of the suspected criminal sending the message. A criminal may get in sessions with as many as possible users which will enforce opening of all their keys. [21]

---

[2] *Early recovery attack* : The computation for the extra *differential-work-factor* can be done off-line before obtaining the warrant.

2.4 Another Approach and Attacks Against It: Message Recovery System

"*Message recovery system*" is a system which does not require the recovery of key. The reference [47,35] introduced the notion of *sharable function* $f(\cdot)$: For all input x each participant outputs a share of the $f(x)$ rather than the actual value of $f(x)$, then any sufficiently large collection of the shares can compute the actual value of $f(x)$. This method is applicable to the message recovery systems by the following procedure [21] :

1. Each recovery agent decrypts the given ciphertext c inside his tamperproof device without revealing the decryption key or cleartext.
2. They reveal the local results of the decryption, i.e. rather than sending a cleartext m, m's share is sent out.
3. Finally, the party combining the shares takes majority and reveals the result m.

Message recovery system is a useful method to limit context and time of wiretapping, prevent the key cloning attack, and so on. But since in the message recovery system the recovery agents can get access to the ciphertext, unlike the key recovery system, there may be a chance for the agents to get information on what message a suspect sends. To solve this problem, we can apply Sakurai's blind decoding protocol [45] into the message recovery, so an authorized agent can ask the message recovery agent to decode the message without revealing what is the decoded plaintext nor learning the recovery agent's secret key. [46] The scheme in [51] and [18] uses the blind decoding technique for the recovery of the short-term secret session key encrypted by user's long-term secret key. An authorized entity can ask an recovery entity to decode the encrypted session key without revealing the session key nor learning the long-term secret key.

But, unfortunately, blind decoding techniques often suffer from the problem of the oracle attack. [2,30,15,38] [3]

As seen above, key escrow cryptography is very difficult to implement in both view of technologies and political debates. Recently, there are some alternatives to key escrow such as "*oblivious key escrow* [12]", "*translucent cryptography [7]*", "*recovering keys in open networks* [20]", and so on.

3 Problem 1 – Mao's VES Is Not Shadow Public-Key Resistant

In [33], Mao presented a VES, Verifiable Escrowed Signature, technique that allows a signer to convince a verifier the validity of a signature without letting

---

[3] *Oracle problem* : The decrypter can be abused as an oracle into releasing useful information for an attacker acting as a requester. The reference [30,15] discussed a countermeasure to the oracle problem by utilizing a *transformable digital signature* : [30] for ElGamal system, [15] for both RSA and ElGamal systems utilizing the transformability of RSA digital signature and ZKIP-digital signature based on commutative random self-reducibility. [39] But the reference [15] is broken by [38].

the verifier see the signature value. This technique can be used in fair exchange of contracts, fair escrow cryptosystems, and so on.

Most of fair escrow cryptosystems (e.g. [34,25,6,40]) establish an escrowed key via real-time communications with each of trustees who must stay on-line to receive messages and verify the correctness of secret sharing. However, Mao's VES can be used to achieve a fair escrow cryptosystem which uses *off-line* trustees. In VES, a single party suffices to interface the users for the off-line trustees, verify the user's messages, and thereby insure that if the messages are later decrypted by the trustees, the user's secret key can be recovered.

3.1 Mao's Verifiable Escrowed Signature Scheme

Let $q = 2r + 1$ and $p = kq + 1$ where, r, q, p are prime. Let further $h \in Z_q^*$ be an element of order r, and $g \in Z_p^*$ be an element of order q. The private/public-key pair of off-line escrow agent, EA, be (S_{EA}, P_{EA}), and that of the prover Alice be (S_A, P_A). Here

$$P_{EA} = h^{S_{EA}} \pmod{q} \quad \text{for } 0 < S_{EA} < r$$

and

$$P_A = g^{S_A} \pmod{p} \quad \text{for } 0 < S_A < q.$$

To sign a message m, Alice chooses k and c such that $0 < c, k < q$ and computes $e = H(g^k \pmod{p}, m)$, $d = c^{-1} \cdot S_A^{-1} \cdot (k - e) \pmod{q}$ and $P' = P_A^c \pmod{p}$. Alice also chooses $0 < K < r$ and makes $A = h^K \pmod{q}$ and $B = P_{EA}^{-K} \cdot c \pmod{q}$. Alice sends to a single party, e.g. Certificate Authority, (d, e, m, P', A, B) and non-interactive zero-knowledge proof [50] for proving (A, B) indeed form a pair of ElGamal encryption of $\log_{P_A} P'$ and the public-key P_{EA} is used for the encryption. A single party can now verify the signature by testing if $H(P'^d \cdot g^e \pmod{p}, m) \overset{?}{=} e$ and the zero-knowledge proof.

To escrow the private-key S_A by off-line, Alice generates different signatures using the different values of c and a same value k. In the case of using two off-line escrow agents, when the two agents later disclose the missing signature parts, the private-key S_A can be recovered by solving a linear congruence.

3.2 Shadow Public-Key Problem and Failsafe Off-Line Key Escrow

Like [34], since in Mao's scheme only Alice has the control over her secret key, it can be abused via the use of shadow public-key cryptosystem. Using a hash function H and Kilian's technique [25], we can easily solve this problem.

To generate Alice's private/public-key pair, Alice firstly chooses $0 < S_A^{(1)} < q$ and computes

$$S_A^{(2)} = H(g^{S_A^{(1)}} \pmod{p}).$$

Now Alice's private-key is

$$S_A = S_A^{(1)} + S_A^{(2)} \pmod{q}$$

and the corresponding public-key is

$$P_A = g^{S_A} \pmod{p}.$$

Then Alice generates her signature as before, and sends to a single party $P_A^{(1)} = g^{S_A^{(1)}}$, (d, e, m, P', A, B) and zero-knowledge proof (here, $P_A^{(1)}$ is sent in secure way). Then a single party first computes

$$P_A = P_A^{(1)} \cdot g^{H(P_A^{(1)})} \pmod{p},$$

verify the signature (d, e, m, P', A, B), and certify on P_A.

Here, Alice can still use her shadow public-key in such a way that Alice firstly makes the pirate(unescrowed) public-key database for $P_A^{(1)}$. Then Bob can obtain a $P_A^{(1)}$ from the pirate database, and, for certification, checks if $P_A \stackrel{?}{=} P_A^{(1)} \cdot g^{H(P_A^{(1)})}$, where P_A and g is obtained from the legal public-key directory. In this case, since the pirate public-key database is only useful when it is available to everyone, the act of running a pirate database is detectable by law-enforcement. This is called a pirate database problem (this is not a kind of subliminal attacks !) and general to other key recovery systems.

Recently, in [41], Park et $al.$ proposed a more efficient verifiable escrowed signature scheme based not on a verifiable encryption of discrete logarithm but on a verifiable encryption of square root. But their scheme has the same weakness as Mao's.

4   Problem 2 – Subliminal Communication Is Easy Using Dawson's Software Key Escrow

Balenson, Ellison, Lipner, and Walker had proposed two protocols to implement a software key escrow system [4]. In [16], Dawson and He pointed out their weakness, and proposed an improvement over two known protocols. Their protocols have some desirable attributes.

1. The negotiation of a common session key before a communication is not necessary. A session key is established during a communication, instead of a $priori$ negotiation.
2. The identities of both the sender and receiver are binded to the content of every transaction. Thus, the LED, Law Enforcement Decryptor, cannot abuse the system.
3. The identity of a user is bound to its software program SP, so that only he himself can use his package. This makes illegal software reproduction hard.
4. The LED, once given a permission from the judge, can only wiretap what he has requested. In other words, he cannot misuse his power.

However, because of the first advantage of protocol, user can abuse the cryptosystem and communicate under the government's very nose without having to worry that it would be detected.

4.1   Abuse of Dawson's Software Key Escrow Encryption Scheme

Now, we demonstrate how A sends a secret message m to B through Dawson's escrow system without being detected even though their keys have been escrowed. Like in [16], A will feed the following inputs to SP_A :

- UID_A, UID_B : The identities of the sender and receiver
- $(KAS1, KAS2)$ and r_A : A's escrowed secret keys and password, for example, the signature of UID_A under A's secret key.
- $(KBP1, KBP2)$: B's certified public-keys
- t : timestamp

In addition, A generates a random number $K1$, computes

$$K2 = m \oplus h(K1), \ K = K1 \oplus K2,$$

chooses an arbitrary message M and feed K, $K1$, $K2$, and M to SP_A.

Then the encryption is done by SP_A as follows :

1. Checks the correctness of the password r_A and the embedded identification number UID_A within the SP_A. This includes signing UID_A and comparing it with r_A.
2. Computes and outputs the following quantities :

$$C = e_K(M)$$
$$a = E_{KAP1}(K1), E_{KAP2}(K2)$$
$$b = E_{KBP1}(K1), E_{KBP2}(K2)$$
$$H = h(UID_A, UID_B, t, K1, K2)$$
$$S = S_A(H)$$
$$LEAF = UID_A, UID_B, KAP1, KAP2, a, b, t, S$$
$$EVS = e_K(UID_A, UID_B, t, H, S).$$

A will send C, $LEAF$ and EVS, Escrow Verification String, to B. Here, we denote a public-key cryptosystem with an encryption/decryption pair by (E, D), the encryption/decryption function of a standard symmetric encryption algorithm by (e, d), a secure one-way hash function by h, and A's signature on a message H by $S_A(H)$.

3. After B obtains $(C, LEAF, EVS)$, he construct the message m simply by decrypting b of $LEAF$ to obtain $K1$ and $K2$, and computing $m = K2 \oplus h(K1)$.

4.2   Vendors' Abuse to Leak Customer's Private-Keys

A manufacturer (we call him M) of a software key escrow encryption scheme can also use the above abuse to leak his customers' private-keys. Let's imagine the following situation : the tempted LED might conspire with a manufacturer

M and have M embed the abuse mentioned above in his implementation. Then he distributes the software products to his customers, e.g. A, B, and everyone else who wants them.

Every time user A sends a message to user B, A will feed his secret keys $(KAS1, KAS2)$ in order to use her own software program SP_A. A's SP_A reads $(KAS1, KAS2)$, and encrypts the message in such a way that $K1 = KAS1 \oplus h(KAP1, KBP1, t)$, $K2 = KAS2 \oplus h(KAP2, KBP2, t)$, and $K = K1 \oplus K2$.

To wiretap the communication from A to B, the LED first obtains a court order from a judge, and presents the court order and $LEAF$ to the two $KEAs$, Key Escrow Agents. Then the first $KEA1$ decrypts the first component of a (using $KAS1$) and sends $K1$ back to the LED. Similarly, the second $KEA2$ decrypts the second component of a (using $KAS2$) and sends $K2$ back to the LED. Now the LED can obtain $KAS1 = K1 \oplus h(KAP1, KBP1, t)$ and $K2 = KAS2 \oplus h(KAP2, KBP2, t)$, and can decrypt everything transmitted by A at any time, whether that transmission was conducted a long time ago or will be conducted in the future. This contradicts the fourth claimed condition of Dawson's scheme.

5   Problem 3 – Lee's Key Escrow System Is Vulnerable to the LEAF Attack and the Squeezing Attack

Lee and Laih [28] introduced a new key escrow system in which, if the investigator cannot recover the message, the receiver cannot either. Lee's key escrow system is described as follows.

Lee's key escrow system. The sender encrypts the session key by TC(Trusted Center)'s public-key PK_{TC} to obtain $E_{PK_{TC}}(KS)$. The message encrypted key is $KS \oplus E_{PK_{TC}}(KS)$, which denotes the corresponding ciphertext of plaintext M to be $E_{KS \oplus E_{PK_{TC}}(KS)}(M)$. Finally, the sender A generates $LEAF$ and forwards it along with ciphertext $E_{KS \oplus E_{PK_{TC}}(KS)}(M)$ to the receiver B. The structure of the $LEAF$ is as follows :

$$LEAF = E_{KF}(E_{PK_{TC}}(KS), E_{PK_B}(D_{SK_A}(KS)), UID_A, UID_B, TS, AC)$$

where $AC = h(KS, UID_A, UID_B, TS)$, KF is a unique family key, $D_{SK_A}(\cdot)$ is A's signature with private-key SK_A, TS is a timestamp, and $h(\cdot)$ is a public one-way hash function.

After receiving the message, the receiver decrypts the $LEAF$ with KF, and obtains the message decryption key $KS \oplus E_{PK_{TC}}(KS)$ from KS and $E_{PK_{TC}}(KS)$. Due to the message is encrypted by $KS \oplus E_{PK_{TC}}(KS)$ instead of KS, if the receiver cannot obtain the exact $E_{PK_{TC}}(KS)$ from $LEAF$, then he cannot recover message M. So if the sender may withhold or not forward the true LEAF to the receiver, the users cannot communicate in privacy.

Weakness of Lee's system. In Lee's system, if user A encrypts the message with key $KS \oplus E_{PK_{TC}}(h(KS))$, and replaces the exact $LEAF$ and AC by $LEAF'$ and AC' such that

$$LEAF' = E_{KF}(E_{PK_{TC}}(h(KS)), E_{PK_B}(D_{SK_A}(KS)), UID_A, UID_B, TS, AC')$$

$$AC' = h(h(KS), UID_A, UID_B, TS),$$

then only the receiver can recover the message but the investigator cannot. Furthermore, assume that user A first opens a private session with user C and obtains $D_{SK_C}(KS')$ from C's $LEAF$. If the user A sends to B the *ciphertext*, $LEAF'$ and AC' such that

$$ciphertext = E_{KS' \oplus E_{PK_{TC}}(KS')}(M)$$

$$LEAF' = E_{KF}(E_{PK_{TC}}(KS'), E_{PK_B}(D_{SK_C}(KS')), UID_C, UID_B, TS, AC')$$

$$AC' = h(KS', UID_C, UID_B, TS),$$

then A can freely frame the honest user C.

6   Problem 4 – Horster's Key Escrow System with Active Investigator Does Not Have Time-Boundness

In [22], Horster *et al.* pointed out the main weaknesses in [8,9] and presented a new software based escrowed key exchange protocol with active investigator. Their scheme has the following properties.

1. Through an ElGamal like signature scheme the user's random parameters can be verified by the network server. Thereby they are related and cannot be substituted randomly.
2. Every user has different unrelated escrowed secret/public-key pairs (x_t, y_t) for every time interval I_t, so the investigator can only decrypt the communication during that time interval.
3. Assume the case in which the investigator has a suspicion and wants to intercept a telephone line, but cannot get a court order for interception in time. In Horster's scheme, as the computations can be done by using the old traffic which can be taped by the network, the investigator can always decrypt old ciphertext if he gets the old secret key from the escrow agency.

In the following we briefly describe the escrowed key exchange protocol given in [22].

Horster's escrowed key exchange protocol. The trusted authority chooses primes p and q with $q|(p-1)$, a generator α of a multiplicative subgroup of order q over Z_p and a collision free hash function h. User Alice(Bob) chooses n secret keys $x_{A(B),i} \in_R Z_q$, computes the related public-keys $y_{A(B),i} = \alpha^{x_{A(B),i}}$ (mod p), and escrows all n secret keys $x_{A(B),i}$ to the key escrow agency.

To exchange a key during the time interval I_t, the network server chooses the messages $m_A, m_B \in_R Z_q$ and submits them to Alice and Bob. Alice(Bob) sends to the network $r_{A(B)}, s_{A(B)}$ where $k_{A(B)} \in_R Z_q$, $r_{A(B)} = \alpha^{k_{A(B)}}$ (mod p) and $s_{A(B)} = x_{A(B),t} \cdot (r_{A(B)} \oplus m_{A(B)}) + k_{A(B)}$ (mod q). Network then verifies the equation $\alpha^{s_{A(B)}} \stackrel{?}{=} y_{A(B),t}^{r_{A(B)} \oplus m_{A(B)}} \cdot r_{A(B)}$ (mod p) and sends $r_{B(A)}$ to Alice(Bob). After receiving $r_{B(A)}$, Alice(Bob) computes the session key $K_{A,B} = h(r_{B(A)}^{x_{A(B),t}} \circ y_{B(A),t}^{k_{A(B)}})$.

When an authorized investigator participates in the protocol during the time interval I_t, he gets the corresponding secret key $x_{A,t}$ from the escrow agency presenting his court order, computes $k_A = s_A - x_{A,t} \cdot (r_A \oplus m_A)$ (mod q), and obtains the session key $K_{A,B}$.

Weakness of Horster's protocol. Consider the case where Alice and Bob use session keys $K_{(A,B),1}, K_{(A,B),2}, \cdots, K_{(A,B),n}$ over a period of time $I_1, I_2, \cdots, I_n$ as in [27]. Using the above escrowed key exchange protocol, they compute t'th session key as

$$K_{(A,B),t} = h(r_{B(A)}^{x_{A(B),t}} \circ y_{B(A),t}^{k_{A(B)}}, K_{(A,B),t-1}) \quad t = 2, 3, \cdots, n.$$

Due to the second property, even though an investigator of the time interval I_t can get the corresponding secret key $x_{A,t}$, it cannot find $K_{(A,B),t}$. Of course the investigator will be able to find the session key $K_{(A,B),t}$ if it knows all previous keys (due to the property 3), but this contradicts the property 2. [4]

7   Problem 5 – Securing Your Bins

In [24], Joye proposed *"garbage-man-in-the-middle attack"*, which relies on the possibility to get access to the "bin" of the recipient.

Joye's garbage-man-in-the-middle attack. If the cryptanalyst intercepts, transforms and re-sends a ciphertext, then the corresponding plaintext will be meaningless when the authorized receiver, Bob, decrypts it. So, Bob will discard it. If the cryptanalyst can get access to this discard, e.g. MS-DOS *undelete* command, he will be able to recover the original plaintext if the transformation is done in a clever way such as Davida's method.

We can also apply Joye's idea to key recovery systems, which result in the framing attack.

Garbage-framing attack. Imagine Alice wants to send a message m to Carol via careless good citizen, Bob. She first computes the ciphertext $c = E_{K_{AC}}(m)$ and post-encrypts c with another session key K_{AB}, and sends it, i.e. $c' = E_{K_{AB}}(E_{K_{AC}}(m))$, to Bob. Here, E is an escrow encryption scheme, and $K_{AB(C)}$ is shared session key between Alice and Bob(Carol). Next, Bob decrypts c', and gets $c = E_{K_{AC}}(m)$. Since the message c has no meaning, Bob discards it. Finally, if Carol can get access to c, then she recovers the message m from c.

In this environment, two collaborating parties, Alice and Carol can frame legitimate user Bob, and fake their own session to look like it is between the legitimate parties.

---

[4] Note that [29] also has the same problem.

8   Problem 6 – Subliminal Channel in Proxy Signatures

In [49], Simmons showed how to embed a subliminal channel in digital signatures created using the DSA signature scheme. But his broadband channel suffers from the necessity that the subliminal receiver know the signer's secret signing key in order to be able to decode subliminal communication.

Let's further consider that a signer Alice asks Bob to carry out signing for her by using Mambo's partial delegation method [31,32,26]. Denote by g a generator of a group of prime order q mod p. The public-key of Alice is $v = g^{s_A}$ (mod p), where $s_A \in Z_q$. An original signer Alice generates $k \in_R Z_q$, $K = g^k$ (mod p), $\sigma = s_A + kK$ (mod q), and gives (σ, K) to a proxy signer, Bob, in a secure manner. [5] Now, if Alice can access to Bob's DSA proxy signatures $r = (g^m \bmod p) \bmod q$, where $m \in Z_q$ is Bob's subliminal message to Alice, and $s = m^{-1}(H(M) - \sigma r) \bmod q$, then the subliminal receiver Alice with σ can decode $m = s^{-1} \cdot (H(M) - \sigma r) \bmod q$. But any other receiver can only verify the proxy signature with a new public value $v' = v_A K^K \bmod p$. Furthermore Alice(or Bob) don't have to know Bob's(or Alice's) secret key s_B(or s_A).

Here, for the same results, we can also use the *delegation by warrant (bearer proxy)* method instead of *partial delegation*. [6] But in key recovery systems, it seem illegal that other entities, except the KRA or CA, generate user's certificates (which results in the pirate public-key database).

9   Conclusions

The concept of key escrow cryptography is very controversial at this stage. We analyzed the claimed requirements of recent key escrow systems and presented how to bypass the protocols and other subliminal attacks that apply to Clipper and recent designs. In order to overcome such attacks we believe that further investigations and technical discussions are needed. We now continue to cryptanalyze recent key escrow proposals and try to make the design principles to construct key recovery systems.

Acknowledgements

We thank the anonymous referees of ISW'99 for comments which improved the presentation of the paper.

---

[5]   For restricting Bob's misbehavior, we can use *partial delegation with warrant* method. [26]

[6]   There are two types of schemes for delegation by warrant approach :

 - (*delegate proxy*) In delegate proxy, Alice signs a document, declaring Bob is designated as a proxy signer, under her secret key by an ordinary signature scheme. The created warrant is given to Bob.
 - (*bearer proxy*) In bearer proxy, a warrant is composed of a message part and an original signer's signature for newly generated public-key. The secret key for a newly generated public-key is given to Bob in a secure way.

References

1. H.Abelson, R.Anderson, S.M.Bellovin, J.Benaloh, M.Blaze, W.Diffie, J.Gilmore, P.G.Neumann, R.L.Rivest, J.I.Schiller, B.Schneier, "The risks of key recovery, key escrow and trusted third-party encryption", May 1997 (rev. 1998).
2. R.Anderson and R.Needham, "Robustness principles for public-key protocols", *Advanced in Cryptology - Crypto'95*, Springer-Verlag, *Lecture Notes in Computer Science*, LNCS 963, 1995, pp.236-247.
3. R.Anderson and M.Roe, "The GCHQ protocol and its problems", *Advanced in Cryptology - Eurocrypt'97*, Springer-Verlag, *Lecture Notes in Computer Science*, LNCS 1233, 1997, pp.134-148.
4. D.M.Balenson, C.M.Ellison, S.B.Lipner, and S.T.Walker, "A new approach to software key escrow encryption", *manuscript*, 1994.
5. M.Bellare and S.Goldwasser, "Encapsulated key escrow", *MIT/LCS Technical Report 688*, April 1996.
6. M.Bellare and S.Goldwasser, "Verifiable partial key escrow", *The 4th ACM Conference on Computer and Communications Security*, 1997.
7. M.Bellare and R.L.Rivest, "Translucent cryptography - An alternative to key escrow, and its implementation via fractional oblivious transfer", *Journal of Cryptology*. 12(2), 1999, pp.117-139.
8. T.Beth, H.-J.Knobloch, M.Otten, G.Simmons, P.Wichmann, "Towards acceptable key escrow systems", *Proc. 2nd ACM Conference on Computer and Communications Security*, Fairfax, Nov. 2-4, 1994, pp.51-58.
9. T.Beth, M.Otten, (ed.), "E.I.S.S.-Workshop on escrowed key cryptography", *E.I.S.S.-Report 94/7*, University of Karlsruhe, June 22-24, 1994, 160 pages.
10. M.Blaze, "Protocol failure in the escrowed encryption standard", *Building in Big Brother : The Cryptographic Policy Debate (Edited by L.J.Hoffman)*, Springer-Verlag, pp.131-146.
11. M.Blaze, "Protocol failure in the escrowed encryption standard", *The 2nd ACM Conference on Computer and Communications Security*, November 1994, pp.59-67.
12. M.Blaze, "Oblivious key escrow", *Cambridge Workshop on Information Hiding*, May 1996.
13. M.Burmester, Y.G.Desmedt, and J.Seberry, "Equitable key escrow with limited time span (or, How to enforce time expiration cryptographically)", *Advanced in Cryptology - Asiacrypt'98*, Springer-Verlag, *Lecture Notes in Computer Science*, LNCS 963, 1514, pp.380-391.
14. CESG, "Securing electronic mail within HMG : Part I. Infrastructure and protocol, Draft C", 21 March 1996, available at *http://www.opengroup.org/public/tech/security/pki/casm/casm.htm*
15. I.Damgard, M.Mambo and E.Okamoto, "Further study on the transformability of digital signatures and the blind decryption", *The 1997 Symposium on Cryptography and Information Security*, SCIS97-33C, 1997.
16. E.Dawson and J.He "Another approach to software key escrow encryption", *First Australasian Conference on Information Security and Privacy, ACISP'96*, Springer-Verlag, *Lecture Notes in Computer Science*, LNCS 1172, 1996, pp.87-95.
17. D.E.Denning and D.K.Branstad, "A taxonomy of key escrow encryption systems", *Communications of the ACM*, 39(3), March 1996, pp.34-40.
18. S.Domyo, U.Hisashi, H.Tsuchiya, K.Toru, T.Tanida, N.Torii, M.Mambo, E.Okamoto, "Development of a Key Recovery System Suitable for the Commercial Use", *The 56th National Convention of the Information Processing Society of Japan*, 6F-05, 1998.

19. The FBI, "Law enforcement REQUIREMENTS for the surveillance of electronic communications", June 1994. (Prepared by the Federal Bureau of Investigations (FBI) in cooperation with federal, state, and local law enforcement members of the National Technical Investigation Association).

20. P.-A.Fouque, G.Poupard and J.Stern, "Recovering keys in open networks", *1999 IEEE-ITW (Information Theory Workshop)*, IEEE, June 1999.

21. Y.Frankel and M.Yung, "Escrow encryption systems visited : Attacks, analysis and designs", *Advanced in Cryptology - Crypto'95*, Springer-Verlag, *Lecture Notes in Computer Science*, LNCS 963, 1995, pp.222-235.

22. P.Horster, M.Michels and H.Petersen, "A new key escrow system with active investigator", *Proc. Securicom*, Paris, La Defense, 8.-9. June, 1995, S.15-28. ; also see *Theoretical Computer Science and Information Security Technical Report TR-95-4-f*, Department of Computerscience, University of Technology Chemnitz-Zwickau.

23. N.Jefferies, C.Mitchell and M.Walker, "A proposed architecture for trusted third party services", in E.Dawson and J.Golic, (eds.), *Cryptography: Policy and Algorithms* - Proceedings : International Conference, Brisbane, Australia, July 1995, Springer-Verlag, *Lecture Notes in Computer Science*, LNCS 1029, Berlin, 1996, pp.98-104.

24. M.Joye and J.-J.Quisquater, "On the importance of securing your bins : The garbage-man-in-the-middle attack", *4th ACM Conference on Computer and Communications Security*, ACM Press, 1997, pp.135-141.

25. J.Kilian and T.Leighton, "Fair cryptosystems, revisited : A rigorous approach to key-escrow", *Advanced in Cryptology - Crypto'95*, Springer-Verlag, *Lecture Notes in Computer Science*, LNCS 963, 1995, pp.208-221.

26. S.J.Kim, S.J.Park, and D.H.Won, "Proxy signatures, revisited", *Proc. of ICICS'97, International Conference on Information and Communications Security*, Springer-Verlag, *Lecture Notes in Computer Science*, LNCS 1334, 1997, pp.223-232.

27. L.R.Knudsen and T.P.Pedersen, "On the difficulty of software key escrow", *Advanced in Cryptology - Eurocrypt'96*, Springer-Verlag, *Lecture Notes in Computer Science*, LNCS 1070, 1996, pp.237-244.

28. Y.-C.Lee and C.-S.Laih, "On the key recovery of the key escrow system", *Thirteenth Annual Computer Security Applications Conference"*, IEEE Computer Society, December 8-12, 1997. pp.216-220.

29. A.K.Lenstra, P.Winkler and Y.Yacobi, "A key escrow system with warrant bounds", *Advanced in Cryptology - Crypto'95*, Springer-Verlag, *Lecture Notes in Computer Science*, LNCS 963, 1995, pp.197-207.

30. M.Mambo, K.Sakurain and E.Okamoto, "How to utilize the transformability of digital signatures for solving the oracle problem", *Advanced in Cryptology - Asiacrypt'96*, Springer-Verlag, *Lecture Notes in Computer Science*, LNCS 1163, 1996, pp.322-333.

31. M.Mambo, K.Usuda, and E.Okamoto, "Proxy signatures : Delegation of the power to sign messages", *IEICE Trans. Fundamentals*, Vol.E79-A/No.9, 1996, pp.1338-1354.

32. M.Mambo, K.Usuda, and E.Okamoto, "Proxy signatures for delegating signing operation", *Proc. Third ACM Conf. on Computer and Communications Security*, 1996, pp.48-57.

33. W.Mao, "Verifiable escrowed signature", *Second Australasian Conference in Information Security and Privacy*, Springer-Verlag, *Lecture Notes in Computer Science*, LNCS 1270, Sydney, July 1997., pp.240-248.

34. S.Micali, "Fair public-key cryptosystems", *Advanced in Cryptology - Crypto'92*, Springer-Verlag, *Lecture Notes in Computer Science*, LNCS 740, 1992, pp.113-138.

35. S.Micali and R.Sidney "A simple method for generating and sharing pseudo-random functions, with Applications to Clipper-like Key Escrow Systems", *Advanced in Cryptology - Crypto'95*, Springer-Verlag, *Lecture Notes in Computer Science*, LNCS 963, 1995, pp.185-196.

36. S.Miyazaki, I.Kuroda and K.Sakurai, "Toward fair international key escrow - An attempt by distributed trusted third agencies with threshold cryptography", *Second International Workshop on Practice and Theory in Public Key Cryptography, PKC'99*, Springer-Verlag, *Lecture Notes in Computer Science*, LNCS 1560, Kamakura, Japan, March 1-3, 1999, pp.171-187.

37. NIST, "Escrow Encryption Standard (EES)", *Federal Information Processing Standards Publication (FIPS PUB) 185"*, 1994.

38. K.Ohta, "Remarks on Blind Decryption", Okamoto, Davida, Mambo (Eds.): *Proc. of ISW'97, Information Security Workshop*, Springer-Verlag, *Lecture Notes in Computer Science*, LNCS 1396, Tatsunokuchi, Ishikawa Japan, September 17-19 1997, pp.273-281.

39. E.Okamoto and K.Ohta, "Divertible zero knowledge interactive proofs and commutative random self-reducibility", *Advanced in Cryptology - Eurocrypt'89*, Springer-Verlag, *Lecture Notes in Computer Science*, LNCS, 1990, pp.134-149.

40. T.Okamoto, "Threshold key-recovery systems for RSA" *IEICE Trans. Fundamentals*, Vol.E82-A/No.1, January 1999, pp.48-54.

41. S.J.Park, S.M.Park, D.H.Won, and D.H.Kim, "An efficient verifiable escrowed signature and its applications", *Journal of the Korean Institute of Information Security and Cryptology*, 8(4), 1998.12., pp.127-138.

42. B.Pfitzmann and M.Waidner, "How to break fraud-detectable key recovery", *ACM Operating Systems Review* 32(1), pp.23-28, January 1998.

43. R.L.Rivest, "Chaffing and winnowing : Confidentiality without encryption", *http://theory.lcs.mit.edu/~rivest/chaffing.txt*, March 18, 1998 (rev. July 1, 1998).

44. R.L.Rivest, A.Shamir and D.A.Wagner, "Time-lock puzzles and timed-release Crypto", March 10, 1996.

45. K.Sakurai, Y.Yamane, "Blind decoding, blind undeniable signatures, and their applications to privacy protection", *Information hiding : first international workshop*, R.J.Anderson, Ed., vol. 1174 of Lecture Notes in Computer Science, Isaac Newton Institute, Cambridge, England, May 1996, Springer-Verlag, Berlin, Germany. ISBN 3-540-61996-8., pp.257-264.

46. K.Sakurai, Y.Yamane, S.Miyazaki and T.Inoue, "A key escrow system with protecting user's privacy by blind decoding", *Proc. of ISW'97, Information Security Workshop*, Springer-Verlag, *Lecture Notes in Computer Science*, LNCS 1396, 1997, pp.147-157.

47. A.De Santis, Y.Desmedt, Y.Frankel and M.Yung, "How to share a function securely", *Proceedings of the 26th Annual Symposium on Theory of Computing*, ACM, 1994, pp.522-533.

48. A.Shamir, "Partial key escrow : A new approach to software key escrow", Presented at *Key escrow conference*, Washington, D.C., September 15, 1995.

49. G.J.Simmons, "Subliminal communication is easy using the DSA", *Advanced in Cryptology - Eurocrypt'93*, Springer-Verlag, *Lecture Notes in Computer Science*, LNCS 765, 1993, pp.T65-T81.

50. M.Stadler, "Publicly verifiable secret sharing", *Advanced in Cryptology - Eurocrypt'96*, Springer-Verlag, *Lecture Notes in Computer Science*, LNCS 1070, 1996, pp.190-199.

51. T.Tanida, H.Tsuchiya, S.Domyo, N.Torii, M.Mambo and E.Okamoto, "Design and Implementation of a Key Recovery System", *The 55th National Convention of the Information Processing Society of Japan*, 2T-01, 1997.

52. Y.Wang, "Abuses of probabilistic encryption schemes", *ELECTRONICS LETTERS*, 16th April 1998, 34(8), pp.753-754.

53. A.Young and M.Yung, "The dark side of black-box cryptography -or- Should we trust capstone ?", *Advanced in Cryptology - Crypto'96*, Springer-Verlag, *Lecture Notes in Computer Science*, LNCS 1109, 1996, pp.89-103.

54. A.Young and M.Yung, "Kleptography : Using cryptography against cryptography", *Advanced in Cryptology - Eurocrypt'97*, Springer-Verlag, *Lecture Notes in Computer Science*, LNCS 1233, 1997, pp.62-74.

55. A.Young and M.Yung, "The prevalence of kleptographic attacks on discrete-log based cryptosystems", *Advanced in Cryptology - Crypto'97*, Springer-Verlag, *Lecture Notes in Computer Science*, LNCS 1294, 1997, pp.264-276.

56. A.Young and M.Yung, "Auto-recoverable auto-certifiable cryptosystems", *Advanced in Cryptology - Eurocrypt'98*, Springer-Verlag, *Lecture Notes in Computer Science*, LNCS 1403, 1998, pp.17-31.

57. A.Young and M.Yung, "Auto-recoverable cryptosystems with faster initialization and the escrow hierarchy", *Second International Workshop on Practice and Theory in Public Key Cryptography, PKC'99*, Springer-Verlag, *Lecture Notes in Computer Science*, LNCS 1560, Kamakura, Japan, March 1-3, 1999, pp.306-314.

58. "Threat and vulnerability model for key recovery (KR)", *http://www.fcw.com/pubs/fcw/1998/0413/web-nsareport-4-14-1998.html*, 2/18/98 NSA, X3.

An Improvement on
a Practical Secret Voting Scheme

Miyako Ohkubo, Fumiaki Miura*, Masayuki Abe, Atsushi Fujioka, and
Tatsuaki Okamoto

NTT Information Sharing Platform Laboratories
Nippon Telegraph and Telephone Corporation
1-1, Hikarinooka, Yokosuka-shi, Kanagawa-ken, 239-0847 Japan
{mookubo,miura,abe,jun,okamoto}@sucaba.isl.ntt.co.jp

Abstract. This paper improves voters' convenience in the secret-ballot
voting scheme offered by Fujioka, Ohta and Okamoto. In their scheme,
all voters have to participate in all stages of the voting scheme; that is,
the registering, the voting and the counting stages. In our scheme voters
do not need to join to the counting stage; hence the voters can *walk away*
once they cast their ballots. This property, realized by the introduction
of distributed talliers, will be beneficial in the practical implementation
of a voting scheme where less round complexity is desired.

1 Introduction

Secret voting is one of the major applications of cryptography. Many studies have
tried to construct secure and efficient schemes. Several approaches can be seen
in the literature: schemes based on general multi-party computation [18,3,10],
schemes that use homomorphic encryption schemes [4,12,27,11,13,26], schemes
constructed over the Mix-net [5,22,28,23,1,20,2], and schemes using blind signatures [8,17].

Each type of the schemes has merits and demerits: Using general multi-party
computation yields a scheme that is inefficient in practice because it requires
vast number of interactions among voters. The schemes based on homomorphic
encryption are efficient in binary choice elections where voters only need to
answer yes or no. However, since the voters have to prove the correctness of their
ballots, they suffer from cumbersome zero-knowledge proofs if such schemes are
used for more complicated elections where voters chose exactly $n/2$ people out
of n candidates for instance. Schemes based on the Mix-net can be used for any
type of election because individual ballots are opened at the end so that anyone
can easily check if the content is correct. The drawback is that it is very costly
to guarantee that every ballot has been opened correctly when the number of
voters is large.

---

* Currently working at NTT Mobile Communications Network Inc.

M. Mambo, Y. Zheng (Eds.): ISW'99, LNCS 1729, pp. 225–234, 1999.
© Springer-Verlag Berlin Heidelberg 1999

Although the schemes that uses blind signatures require somewhat stronger assumptions such as the use of anonymous channels or trusted signing authorities, they are more efficient in terms of computation. One can verify the correctness of the result simply by verifying signatures issued by the signing authorities. Indeed, all reported implementations that were successful in experimental use employ schemes based on blind signatures[1]. In particular, the scheme proposed by Fujioka *et al.* [17] is widely adopted.

A considerable obstacle for the real use of the scheme of [17] is that all voters have to join the ballot counting process. This means that each voter must stay until all other voters complete the vote casting stage. Such an inconvenience for voters can not be allowed in large-scale elections. Furthermore, when it is implemented over a network such as the Internet where establishing stable connections is very expensive, round complexity may degrade the efficiency.

The implementation of [17] introduced in [9] inherits this inconvenience. Although they suggested a slightly modified scheme, it lost fairness; the intermediate result can be known by some authorities if they desire. Another implementation of [17], introduced in [19], eliminates the inconvenience and seem to have no serious drawback in terms of security, but it is computationally inefficient.

The contribution of this paper is to propose a secure and practical blind signature based voting scheme that allows voters to *walk away* once they finish casting their votes. Furthermore, the proposed scheme preserves all the security attributes stated in [17]. In other words, our solution is secure and efficient in the sense that it satisfies the following properties:

- *Completeness* ... All valid votes are counted correctly if all participants follow the protocol.
- *Privacy* .... It is infeasible to associate individual votes and voters.
- *Unreusability* .... Every voter can vote only once.
- *Eligibility* .... Only legitimate users can vote.
- *Fairness* .... No one knows any intermediate result before the deadline has passed.
- *Verifiability* ..... Anyone can verify the correctness of the result.
- *Walk-Away* ..... The voter need not to make any action after voting.

The above properties will be maintained unless the number of colluding authorities exceeds a predetermined threshold: the security parameter. No single authority can violate those security properties and any deviation of the voters will be detected and excluded. Since our goal is to construct a scheme that is useful in practice, this mild level of robustness better suits our purpose than theoretical robustness which means strength against any form of collusion even if it is unlikely to happen in reality.

This paper is organized as follows: **Section 2** describes our model and notations. **Section 3** reviews the scheme of [17] and its two implementations. The proposed scheme is introduced in **Section 4** and we discuss implementation issues in **Section 5**. Concluding remarks appears in **Section 6**.

---

[1] See http://www.ccrc.wustl.edu/~lorracks/sensus/hotlist.html.

2   Model and Notations

2.1   Model

Participants: The participants of the proposed voting scheme are N voters, an administrator, n talliers, and a verifier. The verifier can be any party that want to assure the correctness of the result. The computational power of all those participants is assumed to be polynomial against a predetermined security parameter. Among those participants, there may be adversaries who deviate from the protocol. Their primary purpose is to produce incorrect result, disrupt the voting, or to violate voter privacy. It is assumed that there are up to t active adversaries among n talliers. In the basic scheme introduced in **Section 4**, the administrator is assumed to be trustworthy for the sake of simplicity. We address the deviation of the administrator in **Section 5**.

Communication Channels: We assume the use of a bulletin board where anybody can post without identifying themselves, and nobody can erase or overwrite the data once written. An equivalent scenario was also used in [5,7,21]. Such a primitive may be implemented as a publicly monitored web page where voters post from public terminals. Another possibility for implementing the anonymous access to the bulletin board is to use a Mix-net [5]. This will be discussed in **Section 5**.

Cryptographic Tools: Our scheme uses some cryptographic primitives such as a threshold encryption scheme [15,13], a digital signature scheme [24,25], and a blind signature scheme [6,14]. Since any secure implementation of those primitives suits our scheme, we will describe our scheme using a high level description without specifying which particular schemes are used.

2.2   Notations

The bellow terms have the following meanings unless otherwise specified.

V_i: Voter i
A: Administrator
T: Talliers (T_j denotes the tallier j)
$\xi_T(\cdot)$:   Talliers T's threshold encryption scheme
$\sigma_i(\cdot)$:   Voter V_i's signature scheme
$\sigma_A(\cdot)$:   Administrator's signature scheme
$\chi_A(\cdot)$:   Blinding procedure
$\delta_A(\cdot)$:   Unblinding procedure corresponding to $\chi_A(\cdot)$
ID_i: Voter V_i's identification
v_i: Vote of voter V_i

3   The FOO92 Scheme and Its Implementations

3.1   The FOO92 Scheme

In this subsection, we review the secret voting scheme proposed by Fujioka *et al.* [17]. Their scheme is one of the standard schemes based on blind signatures, and it uses the bit-commitment scheme to realize the fairness property.

The participants of their scheme are voters, an administrator, and *single* tallier. The scheme uses a bit-commitment scheme $\xi(v, k)$ for message v and key k.

- o REGISTERING stage
  - Voter V_i selects vote v_i and commits it to x_i as $x_i = \xi(v_i, k_i)$ using randomly chosen key k_i. V_i then blinds x_i by computing $e_i = \chi(x_i, r_i)$ where r_i is a randomly chosen blinding factor, and signs e_i as $s_i = \sigma_i(e_i)$. V_i sends $\langle ID_i, e_i, s_i \rangle$ to Administrator A.
  - Administrator A checks if
    - $*$ s_i is a valid signature of e_i,
    - $*$ ID_i is registered, i.e., V_i has the right to vote, and
    - $*$ V_i has never applied for a signature.

    If all checks pass, A signs $d_i = \sigma_A(e_i)$ to e_i. A then sends d_i back to V_i.
  - V_i unblinds d_i to obtain the signature y_i as $y_i = \delta(d_i, r_i)$. V_i checks that y_i is a valid signature of the administrator for message x_i. If the check fails, V_i claims so by publishing $\langle x_i, y_i \rangle$.
  - At the end of the REGISTERING stage, A announces the number of voters who were given the administrator's signature, and publishes the list that contains $\langle ID_i, e_i, s_i \rangle$.
- o VOTING stage
  - Voter V_i sends $\langle x_i, y_i \rangle$ to Tallier T via an anonymous channel.
  - Tallier T checks the signature y_i of the ballot x_i using the administrator's verification key. If the check succeeds, T enters $\langle x_i, y_i \rangle$ onto a list with a number l. At the end of the VOTING stage, T announces publishes the list containing $\langle l, x_i, y_i \rangle$.
- o COUNTING stage
  - V_i checks if
    - $*$ the number of ballots on list is equal to the number of voters, and
    - $*$ V_i's ballot is listed on the list.

    If both checks pass, V_i sends k_i with the number l, i.e., $\langle l, k_i \rangle$ to Tallier T via the anonymous channel.
  - Tallier T opens the commitment of the ballot x_i, retrieves the vote v_i (or T adds k_i and v_i to the list), and checks that v_i is valid voting. T counts the voting and announces the voting results.

Privacy and unreusability are consistent because we use blind signatures and anonymous channels. Furthermore, the scheme ensures the fairness since the voters open their votes after all ballots are published. However, the scheme doesn't satisfy the walk-away condition because voters themselves must act again after voting.

3.2   Example 1: Sensus

Cranor and Cytron designed and implemented a secret voting system, called *Sensus*, based on the FOO92 scheme [9].

Sensus uses the RSA scheme [24] as both the cryptosystem and the signature scheme. They implemented it in C and Perl using the RSAREF library, the modules run on a machine with a Web server that supports CGI scripts.

It is assumed that voters communicate with the administrator and the tallier via a special anonymous channel that can transmit messages to both sides. This channel is outside the implementation.

All necessary public keys and private keys are generated and stored in a secure way. Every potential voter V_i registers his identity ID_i.

Since the REGISTERING stage is almost same as the FOO92 scheme, let us start with the VOTING stage.

- o VOTING stage
  - Voter V_i encrypts the ballot with A's signature, and sends the ciphertext to Tallier T.
  - Tallier T decrypts it, and checks that the ballot has A's valid signature. T then, signs it, and returns its signature with a randomly assigned receipt number.
- o COUNTING stage
  - Voter V_i checks T's signature, and sends the commitment key with the receipt number to Tallier T.
  - Tallier T opens the ballot with the received key to retrieve the vote, counts the voting, and announces the voting result.

It is clear that Sensus is a direct implementation of the FOO92 scheme (without the anonymous channel). So fairness is assured but walk-away is not realized as in the FOO92 scheme. In [9], it is suggested that the commitment key can be sent to Tallier just after Voter receives the receipt in the VOTING stage. This modification, however, drops fairness to realize walk-away.

3.3   Example 2: E-Vox

E-Vox is an implementation of the FOO92 scheme by Herschberg and his colleagues [19]. There is an extra participant, called **Anonymizer**, who takes encrypted messages and outputs decrypted messages in random order. That is, Anonymizer can be regarded as a single-server Mix-net.

In the following description, "a private channel" is established over a public communication channel by using public key encryption. Since encrypted messages can be retrieved only by the designated receiver, receiver authentication is provided implicitly.

All necessary public keys and private keys are generated and stored in a secure way. Every potential voter V_i registers his identity ID_i and password PWD_i at the registration desk.

Since the REGISTERING stage is the same as the FOO92 scheme, let us start with the VOTING stage.

o VOTING stage
  - V_i encrypts the vote, the commitment, and the unblinded signature with the tallier's public key. Moreover, the result is encrypted with the anonymizer's public key. The doubly encrypted ciphertext is then sent to the anonymizer. (Note that the vote and the commitment are in the same ciphertext.)
o COUNTING stage
  - Anonymizer decrypts all the ciphertexts and obtains the ciphertext associated with the tallier's public key. It sends the ciphertexts to the tallier in random order.
  - The tallier decrypts each ciphertext. It checks that the commitment is correct with regard to the vote, and verifies the administrator's signature on the commitment. The tallier publishes all the results of decryption.

In this implementation, commitment fails in its original task because a voter sends his vote and commitment at the same time.

4   Proposed Voting Scheme

The participants of the proposed voting scheme are voters, an administrator, and *several* talliers. Our scheme uses a $(t+1, n)$-threshold encryption scheme to ensure the fairness property instead of a bit-commitment.

First we outline the proposed scheme. The scheme consists of the following stages executed by the voters, the administrator, and the talliers. Roughly speaking, the voter prepares a ballot, get an authorization, and votes anonymously. Administrator gives authorization to eligible voters, and talliers cooperatively open the ballots.

By communicating with the administrator, the voter gets authorization and then acts anonymously. The blinding signature technique provides the separation between the identification and anonymous communication.

Here we explain the scheme in detail. (See notations in **Subsection 2.2**.)

o REGISTERING stage
  Step 1: Voter V_i selects vote v_i and encrypts v_i with talliers' public key (of the threshold encryption scheme) as $x_i = \xi_T(v_i)$. V_i blinds x_i as $e_i = \chi(x_i, r_i)$ where r_i is a randomly chosen blinding factor. V_i signs $s_i = \sigma_i(e_i)$ to e_i and sends $\langle ID_i, e_i, s_i \rangle$ to Administrator A.
  Step 2: Administrator A checks that voter V_i has the right to vote. If V_i doesn't have the right, A rejects the authorization. A checks that V_i has not already applied for a signature. If V_i has already applied, A rejects the authorization. A checks the signature s_i of message e_i. If they are valid, then A signs $d_i = \sigma_A(e_i)$ to e_i and sends d_i as A's certificate to V_i. At the end of the REGISTERING stage, A announces the number of voters who were given the administrator's signature, and publishes a list containing $\langle ID_i, e_i, s_i \rangle$.

Step 3: Voter V_i retrieves the desired signature y_i of ballot x_i by $y_i = \delta(d_i, r_i)$. V_i checks that y_i is the administrator's signature of x_i. If the check fails, V_i claims it by showing that $\langle x_i, y_i \rangle$ is invalid.

○ VOTING stage

Step 1: V_i sends $\langle x_i, y_i \rangle$ to the bulletin board through the anonymous channel.

○ COUNTING stage

Step 1: Each Tallier T_j checks the signature y_i of ballot x_i using the administrator's verification key. If the check fails and all talliers agree on this, $\langle x_i, y_i \rangle$ is invalidated and excluded hereafter.

Step 2: All Talliers T_j cooperatively decrypt ballot x_i and retrieve vote v_i in communication via the bulletin board. The talliers publishes the voting results.

Step 3: Verifier checks that the number of ballots (including excluded ones) on the list equals the number of voters and that all talliers act correctly. If the checks fails, Verifier claims this.

The proposed scheme inherits most of the security properties of the FOO92 scheme. The major differences are the number of talliers and the use of the threshold encryption scheme instead of bit-commitment. So this means that the completeness condition is preserved. Unreusability and eligibility also hold under the assumptions that no voter can break the blind signature scheme and no one can break the ordinary digital signature scheme, respectively.

The relation between the voter's identity ID_i and ballot x_i is hidden by the blind signature scheme. Ballot x_i is sent through the anonymous channel, so no one can violate voter privacy. It is unconditionally secure against vote tracing. It ensures that privacy is maintained.

The most important point is that the walk-away property is satisfied since the voters may walk away after casting their votes. They need not to send any information to open their ballots because they encrypt their votes with the talliers' cryptosystem. Furthermore, fairness is also assured because curious talliers can not decrypt the ballots during VOTING stage as long as they are less than threshold $t + 1$.

Verifiability is assured under the assumption that there is no conspiracy.

5   Extension

Implementing anonymous channels The largest obstacle for implementing the proposed scheme is anonymous vote casting. A practical solution to this problem is to use a Mix-net as an anonymous channel. Although a universally verifiable Mix-net can be used for providing reliable proof of correctness, a simple (hence non-verifiable) Mix-net is a practical choice given its excellent efficiency. The following is an example of how to integrate a simple Mix-net into our scheme. The underlying idea is to *back-trace* the Mix-net if invalid signatures are found in the list of opened ballots.

o VOTING stage

Step 1: Each voter encrypts $\langle x_i, y_i \rangle$ with the encryption key(s) of the Mix-net and sends the resulted ciphertext, say c_i, to the bulletin board via a *sender-authenticated public channel*.

o COUNTING stage

Step 1: The Mix-servers decrypts the list of c_i and outputs the list of $\langle x_i, y_i \rangle$ in random order.

Step 2: Talliers T check the signature y_i of x_i. If it fails, T claims so by publishing $\langle x_i, y_i \rangle$. If more than t talliers claim about the same $\langle x_i, y_i \rangle$, the Mix-servers have to reveal the corresponding c_i and prove in zero-knowledge that $\langle x_i, y_i \rangle$ is the correct result of decryption of c_i (thus back-tracing). Each tallier check the proofs issued by each Mix-server. If the check fails, the Mix-server that issued wrong proof is disqualified. If all proofs are okay, it means that the user cast an invalid vote. Thus, the vote is excluded from further steps of the counting stage.

Step 3: After excluding the invalid results of Mix processing, the talliers perform decryption and counting as described in the previous section.

If privacy is to be maintained for the voters whose votes were broken by the deviating Mix-server, those votes must not be opened. So it is assumed that the result of the election will not be affected by those unopened votes.

In a realistic use of our voting scheme, the Mix-servers (i.e., its supervisors) will be carefully appointed. Moreover the fact that a deviating server is always identified will act as a deterrent to malicious behavior.

Tolerating administrator deviation In our basic scheme, the administrator can illegally perform the voting stage by pretending to be a user who performed the registration stage but is not likely to perform the voting stage. Such behavior can be prevented if the anonymous access to the bulletin board is implemented with a Mix-net because Mix-net collects inputs via authenticated channels.

6 Conclusion

This paper proposed a practical secret voting scheme for large scale elections. The scheme ensures voter privacy and prevents any disruption by voters or the administrator. Furthermore, voting fairness and the walk-away property are ensured.

References

1. M. Abe, "Universally Verifiable Mix-Net with Verification Work Independent of The Number Of Mix-Servers", in *Advances in Cryptology — EUROCRYPT '98*, Lecture Notes in Computer Science 1403, Springer–Verlag, Berlin, pp.437–447 (1998).

2. M. Abe, "A Mix-network on Permutation Networks", IEICE Technical Report (May, 1999).
3. M. Ben-Or, S. Goldwasser, and A. Wigderson, "Completeness Theorems for Non-Cryptographic Fault-Tolerant Distributed Computation", Proceedings of the 20th Annual ACM Symposium on Theory of Computing, pp.1–10 (May, 1988).
4. J. Benaloh and M. Yung, "Distributing the Power of a Government to Enhance the Privacy of Votes", Proceedings of the 5th ACM Symposium on Principles of Distributed Computing, pp.52–62 (Aug., 1986).
5. D. L. Chaum, "Untraceable Electronic Mail, Return Addresses, and Digital Pseudonyms", *Communications of the ACM*, Vol.24, No.2, pp.84–88 (Feb., 1981).
6. D. Chaum, "Security without Identification: Transaction systems to Make Big Brother Obsolete", *Communications of the ACM*, Vol.28, No.10, pp.1030–1044 (Oct., 1985).
7. D. Chaum, "The Dining Cryptographers Problem: Unconditional Sender and Recipient Untraceability", *Journal of Cryptology*, Vol.1, No.1, pp.65–75 (1988).
8. D. Chaum, "Elections with Unconditionally-Secret Ballots and Disruption Equivalent to Breaking RSA", in *Advances in Cryptology — EUROCRYPT '88*, Lecture Notes in Computer Science 330, Springer–Verlag, Berlin, pp.177–182 (1988).
9. L. F.. Cranor and R. K. Cytron, "Design and Implementation of a Practical Security-Conscious Electronic Polling System", WUCS-96-02, Deportment of Computer Science, Washington University, St. Louis (Jan., 1996).
10. D. Chaum, C. Crépeau, and I. Damgård, "Multiparty Unconditionally Secure Protocols", Proceedings of the 20th Annual ACM Symposium on Theory of Computing, pp.11–19 (May, 1988).
11. R. Cramer, M. Franklin, B. Schoenmakers, and M. Yung, "Multi-Authority Secret-Ballot Elections with Linear Work", in *Advances in Cryptology — EUROCRYPT '96*, Lecture Notes in Computer Science 1070, Springer–Verlag, Berlin, pp.72–83 (1996).
12. J. Cohen and M. Fisher, "A Robust and Verifiable Cryptographically Secure Election Scheme", 26th Annual Symposium on Foundations of Computer Science, IEEE, pp.372–382 (Oct., 1985).
13. R. Cramer, R. Gennaro, and B. Schoenmakers, "A Secure and Optimally Efficient Multi-Authority Election Scheme", in *Advances in Cryptology — EUROCRYPT '97*, Lecture Notes in Computer Science 1233, Springer–Verlag, Berlin, pp.103–118 (1997).
14. J. Camenisch, J. Pireteau, and M. Stadler, "Blind Signatures based on the Discrete Logarithm Problem", in *Advances in Cryptology — EUROCRYPT '94*, Lecture Notes in Computer Science 950, Springer–Verlag, Berlin, pp.428–432 (?1994?).
15. Y. G. Desmedt and Y. Frankel, "Threshold Cryptosystems", in *Advances in Cryptology — CRYPTO '89*, Lecture Notes in Computer Science 435, Springer–Verlag, Berlin, pp.307–315 (1990).
16. W. Diffie and M. E. Hellman, "New Directions in Cryptography", *IEEE Transactions on Information Theory*, Vol.IT-22, No.6, pp.644–654 (Nov., 1976).
17. A. Fujioka, T. Okamoto, and K. Ohta. "A Practical Secret Voting Scheme for Large Scale Elections", in *Advances in Cryptology — AUSCRYPT '92*, Lecture Notes in Computer Science 718, Springer–Verlag, Berlin, pp.244–251 (1993).
18. O. Goldreich, S. Micali, and A. Wigderson, "How to Play Any Mental Game or a Completeness Theorem for Protocols with Honest Majority", Proceedings of the 19th Annual ACM Symposium on Theory of Computing, pp.218–229 (May, 1987).

19. M. A. Herschberg, "Secure Electronic Voting Over the World Wide Web", Master Thesis in Electrical Engineering and Computer Science, Massachusetts Institute of Technology (1997).

20. M. Jakobsson, "A Practical Mix", in *Advances in Cryptology — EUROCRYPT '98*, Lecture Notes in Computer Science 1403, Springer–Verlag, Berlin, pp.448–461 (1998).

21. A. Pfitzmann, "A Switched/Broadcast ISDN to Decrease User Observability", 1984 International Zurich Seminar on Digital Communications, IEEE, pp.183–190 (Mar., 1984).

22. C. Park, K. Itoh, and K. Kurosawa, "Efficient Anonymous Channel and All/Nothing Election Scheme", in *Advances in Cryptology — EUROCRYPT '93*, Lecture Notes in Computer Science 765, Springer–Verlag, Berlin, pp.248–259 (1994).

23. W. Ogata, K. Kurosawa, K. Sako, and K. Takatani, "Fault Tolerant Anonymous Channel", in *ICICS98*, Lecture Notes in Computer Science 1334, Springer–Verlag, Berlin, pp.440–444 (1998).

24. R. Rivest, A. Shamir, and L. Adleman, "A Method for Obtaining Digital Signatures and Public-Key Cryptosystems," *Communications of the ACM*, Vol. 21, No. 2, pp. 120–126 (Feb., 1978).

25. C. P. Schnorr, "Efficient Signature Generation for Smart Cards", *Journal of Cryptology*, Vol.3, No.4, pp.239–252 (1991).

26. B. Schoenmakers, "A Simple Publicly Verifiable Secret Sharing Scheme and its Application to Electronic Voting", Crypto '99 (To appear).

27. K. Sako and J. Kilian, "Secure Voting using Partially Compatible Homomorphisms", in *Advances in Cryptology — CRYPTO '94*, Lecture Notes in Computer Science 839, Springer–Verlag, Berlin, pp.411–424 (1994).

28. K. Sako and J. Kilian, "Receipt-Free Mix-Type Voting Scheme — A Practical Solution to the Implementation of a Voting Booth —", in *Advances in Cryptology — EUROCRYPT '95*, Lecture Notes in Computer Science 921, Springer–Verlag, Berlin, pp.393–403 (1995).

Undeniable Confirmer Signature

Khanh Nguyen, Yi Mu, and Vijay Varadharajan

School of Computing and Information Technology
University of Western Sydney, Nepean
P.O.BOX 10 Kingswood NSW 2747, Australia
{qnguyen,yimu,vijay}@cit.nepean.uws.edu.au

Abstract. In undeniable signature, a signature can only be verified with cooperation of the signer. If the signer refuses to cooperate, it is infeasible to check the validity of a signature. This problem is eliminated in confirmer signature schemes where the verification capacity is given to a confirmer rather than the signer. In this paper, we present a variation of confirmer signature, called undeniable confirmer signature in that both the signer and a confirmer can verify the validity of a signature. The scheme provides a better flexibility for the signer and the user as well as reduces the involvement of designated confirmers, who are usually *trusted* in practice. Furthermore, we show that our scheme is divertible, i.e., our signature can be blindly issued. This is essential in some applications such as subscription payment system, which is also shown.

1 Introduction

Undeniable signatures, initially introduced by Chaum and van Antwerpen[4], provide the signer with an additional control over her signatures such that the signer's cooperation is required in order to verify an undeniable signature. Also if the signer issues a signature, it is infeasible for her to deny the ownership of the signature. Undeniable signatures have many applications such as sealed-bid auction, software copyright protection amongst many others. Many practical realizations of undeniable signature schemes have been proposed in literature[2,5,9,11,14].

Unfortunately, in undeniable signature schemes, it is infeasible to check the validity of a signature if the signer refuses to cooperate. As signatures often represent signer's commitments, it is always incentives for the signer to do so. A more desired solution is confirmer signatures, introduced by Chaum[6]. In this model, the ability to verify signatures is delegated to an additional entity, called the confirmer. The confirmer is able to either confirm or disavow the validity of a signature but hee is not able to forge any signature. Since its invention, several concrete realizations of confirmer signatures were presented[16,15]. Particularly, the schemes of [15] are efficient and provably secure under the random oracle paradigm[1].

In most confirmer signature schemes, a signer cannot confirm the validity of her signature. If the confirmer refuses to cooperate, it is infeasible to check the

M. Mambo, Y. Zheng (Eds.): ISW'99, LNCS 1729, pp. 235–246, 1999.
© Springer-Verlag Berlin Heidelberg 1999

validity of a signature. In practice, a trusted party would play the role of the confirmer, thus reducing the involvement of any such party is highly desirable from both technical and economical grounds. This is achieved if a signature can be verified with either the cooperation of the signer or a confirmer. Then the user can ask the signer to verify a signature. As a safe guard, the confirmer can also verify a signature if the signer refuses to cooperate.

In this paper, we propose a signature scheme, called undeniable confirmer signature that offers both signer's and confirmer's signature verification capacity. In addition our scheme is divertible, i.e., our undeniable confirmer signatures can be blindly generated. Blind undeniable confirmer signatures are useful in many applications such as subscription payment systems for large network of services where the privacy of network users should be protected while black market selling of subscription tokens must be prevented. There exists only another scheme[8] that achieved the same feature as our undeniable confirmer signature. Though it was proposed in a different context of fair exchange of digital signatures.

The construction of this paper is as follows. Section 2 defines our model of an undeniable confirmer signature scheme. Section 3 introduces our construction of an undeniable confirmer signature scheme. Section 4 proves its security. Section 5 introduces the blind version and its application to subscription-based payment system.

2   A Model of Undeniable Confirmer Signature

In this section, we provide a general model for undeniable confirmer signatures. The model is based on the model of confirmer signatures from [15], which adopts a similar approach to the model of undeniable signatures from [9]. It offers a uniform definition for cryptographic protocols using the standard concepts of interactive Turing machine, interaction proof systems and zero-knowledge. The reader is referred to [12] for details of these terminologies.

For simplicity, we use S to denote a signer, C a confirmer, V a verifier. A undeniable confirmer signature scheme consists of the following algorithms and protocols:

- Two key generator algorithms GEN_S and GEN_C which receive 1^l as input, where l is the security parameter, and respectively output two tuples (S_s, P_s) and (S_c, P_c). Algorithm GEN_S is executed by S, GEN_C by C. (S_s, P_s) and (S_c, P_c) are S's, C's secret/public key pairs respectively. The secret key S_s is used to generate signatures. Also S_s, S_c are used respectively by the signer and the confirmer in confirmation/disavowal protocols.
- A probabilistic polynomial signature algorithm $SIGN$ which receives a secret key S_s and a message m and outputs a signature σ.
- An interactive signature verification protocol $\langle C_{Ver}, V_{Ver} \rangle$. This is a pair of interactive polynomial time Turing machine between the confirmer and the verifier:

$$\langle C_{Ver}(S_c), V_{Ver}() \rangle (m, \sigma, P_s, P_c) \to v$$

The common input consists of a message m, a signature σ, two public keys P_s and P_c. The confirmer possesses S_c as private input. The protocol is designed to return as the verifier's output a boolean value v. If the output is 1, it indicates that σ is a valid signature on the message m and 0 otherwise.

- An interactive signature verification protocol $\langle S_{Ver}, V_{Ver} \rangle$. This is a pair of interactive polynomial time Turing machine between the signer and the verifier:

$$\langle S_{Ver}(S_s), V_{Ver}() \rangle (m, \sigma, P_s, P_c) \to v$$

The common input consists of a message m, a signature σ, two public keys P_s and P_c. The signer possesses S_s as private input. The protocol is designed to return as the verifier's output a boolean value v. If the output is 1, it indicates that σ is a valid signature on the message m and 0 otherwise.

In order for the scheme to make sense, the following security requirements must be met by the protocol's components.

- *Indistinguishability of signatures:* A signature simulator $SIGN_{sim}$. This is a probabilistic polynomial time algorithm which receives a message m, the public keys P_s, P_c as input and outputs an element, called a simulated signature in signature space. This simulated signature is indistinguishable from real signatures for anyone who knows only public information. Given a message and a purported signature, one cannot determine on her own whether the signature is valid.
- *Unforgeability of signatures:* There exists no polynomial time algorithm, which receives the signer's public P_s, both of the confirmer's secret/public keys S_c, P_c and given access to signing oracle $SIGN$, outputs with non-negligible probability a message-signature pair (m', σ') that has not been generated by $SIGN$.
- *Correctness of Verification:* Regardless whether the signer or the confirmer is involved, verification protocols are consistent. Except with negligible probability, a verification protocol returns 1 as the verifier's output if the message signature pair (m, σ) is valid or 0 if (m, σ) is invalid.

3   Preliminaries

In this section, we define some notations and present Schnorr signature scheme and Chaum-Petersen signature of equality of discrete logarithms. These two protocols are special cases of signatures of knowledge[3]. The latter also the non-interactive version of the well-known Chaum-Petersen proof of equality of discrete logarithms. Both these protocols can be proven secure in random oracle model.

3.1   Notation

The symbol $\|$ denotes the concatenation of two binary strings. Also Let p, q be large prime numbers such that q divides $p - 1$. Further let g denote a generate of subgroup G of Z_p^* of order q. Finally, we assume a collision resistant hash function $\mathcal{H}(: \{0,1\}^* \to Z_q^*$ $(k > 160)$.

3.2 Schnorr Signature Scheme

Definition 1. *Given* $y = g^x \bmod p$, *a Schnorr signature on the message* m *verifiable using the public key* (g, y) *is the pair* $(u, v) \in Z_q^* \times Z_q^*$ *satisfying* $u = \mathcal{H}(m\|y\|g\|g^u y^v)$.

Such a signature can be computed if the secret key x is known by choosing $r \in_R Z_q^*$ and computing:

$$u = \mathcal{H}(m\|y\|g\|g^r) \text{ and } v = r - ux \bmod q.$$

For simplicity, we use $\mathcal{S}(x, y)(m)$ to denote a Schnorr signature on the message m generated with the secret key x and verifiable with the public key y.

3.3 Chaum-Petersen Signature of Equality of Discrete Logarithms

Definition 2. *Given* $y_1 = g_1^x$ *and* $y_2 = g_2^x$, *a Chaum-Petersen signature of equality of discrete logarithms of* y_1 *and* y_2 *to the base* g_1 *and* g_2 *on the message* m *is the pair* $(u, v) \in Z_q^* \times Z_q^*$ *satisfying* $u = \mathcal{H}(m\|y_1\|y_2\|g_1\|g_2\|g_1^u y_1^v\|g_2^u y_2^v)$.

Under random oracle model, such a signature can be constructed if the secret key x satisfying $y_1 = g_1^x$ and $y_2 = g_2^x$, is known. The signature is then computed by choosing $r \in_R Z_q^*$, computing:

$$u = \mathcal{H}(m\|y_1\|y_2\|g_1\|g_2\|g_1^u y_1^v\|g_2^u y_2^v) \text{ and } v = r - ux \bmod q.$$

For simplicity, we use $\mathcal{CP}(x, y_1, y_2, g_1, g_2)(m)$ to denote a Schnorr signature on the message m generated with the secret key x showing the equality of discrete logarithms of y_1 and y_2 to the base g_1 and g_2 respectively.

3.4 Fujioka-Okamoto-Ohta Interactive Bi-proof of Equality

A bi-proof of equality $\log_{g_1}(y_1) \equiv \log_{g_2}(y_2)$ is a protocol that either proves $\log_{g_1}(y_1) \equiv \log_{g_2}(y_2)$ or proves $\log_{g_1}(y_1) \neq \log_{g_2}(y_2)$. We present the protocol of Fujioka et. al. here to (dis)prove the equality of discrete logarithm of y_1 and y_2 to the base g_1 and g_2 respectively. The protocol is as follows:

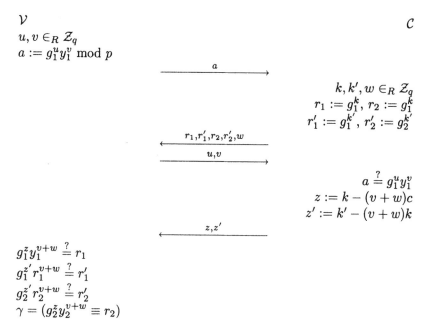

In this protocol, at each verification $L \stackrel{?}{=} R$, the active party would halt the protocol if $L \neq R$. At the end of the protocol, the verifier output is γ. The proof returns 1 if $\log_{g_1}(y_1) = \log_{g_2}(y_2)$ and 0 otherwise. We denote this protocol as $Bi - Proof[\log_{g_1}(y_1) \equiv \log_{g_2}(y_2)]$.

4   The Construction

The undeniable confirmer signature scheme then works as follows:

- *Key Generation:* The signer chooses $s \in_R Z_q$ and forms her secret/public key pair (S_s, P_s) as $S_s = s$ and $P_s = g^s \bmod p$. Similarly the confirmer chooses $c \in_R Z_q$ and computes his secret/public key pair (S_c, P_c) as $S_c = c$ and $P_c = g^c \bmod p$.
- *Signature Generation:* To generate a signature σ on a message m, the signer S chooses $r \in_R Z_q$, generates: $\alpha := g^r$, $\alpha_s := P_s^r$, $\alpha_{s+c} := (P_s P_c)^r$, $g_s := P_s$ and $g_{s+c} := P_s P_c$. S then computes $\sigma_1 = CP(r, \alpha, \alpha_{s+c}, g, g_{s+c})(m)$ and $\sigma_2 = S(sr, g, \alpha_s)(\sigma_1)$. The signature σ then is $\sigma = (\sigma_1, \sigma_2)$.
- *Signature Verification:* The verifier first checks the validity of (σ_1, σ_2), i.e., σ_1 is a valid Chaum-Petersen signature of equality of discrete logarithms on m and σ_2 is a valid Schnorr signature on σ_1. The verifier halts if any check fails. Otherwise, the verifier verifies the signature as follows:
  - With the signer: The verifier's output v of

$$\langle S_{Ver}(S_s), V_{Ver}()\rangle(m, \sigma, P_s, P_c)$$

is computed as $v = Bi - Proof[\log_\alpha(\alpha_s) \equiv \log_g(g_s)]$. In this protocol, the signer assumes the role of the prover.

- With a confirmer: The verifier's output v of

$$\langle C_{Ver}(S_c), V_{Ver}() \rangle (m, \sigma, P_s, P_c)$$

is computed as $v = Bi - Proof[\log_g(g_c) \equiv \log_\alpha(\alpha_c)]$. In this bi-proof protocol, the confirmer acts as the prover and $\alpha_c = \alpha_{s+c}/\alpha_s$.

In both verifications, the verifier accepts the signature if and only if $v = 1$.

4.1   Intuition Explained

In this construction, the signer secret key is s, public key is g_s and the confirmer secret/public key is c and g_c respectively. The value g_{s+c} is computed as $g_{s+c} = g_s g_c$. An undeniable confirmer signature σ consists of two signatures σ_1, σ_2. The former σ_1 is a Chaum-Petersen signature generated with secret key $r_1 = r$, verifiable with public key $\alpha = g^r$ and $\alpha_{s+c} = g_{s+c}^r$. The latter σ_2 is a Schnorr signature, generated with the secret key $r_2 = rs$, verifiable with the public key $\alpha_s = g_s^r$.

Intuitively, a signature is a demonstration of the knowledge of the secret key. So if one can generates σ_1, σ_2, one must possess the knowledge of r_1 and r_2. If one can demonstrate that $r_2 = r_1 s$, it indicates that the the signature generator knows s which implies that the signature is valid.

Proving $r_2 = r_1 s$ can be done in two different ways. One is to prove that $\log_g(g_s) \equiv \log_\alpha(\alpha_s)$. The other is to prove that $\log_g(g_c) = \log_\alpha(\alpha_{s+c}/\alpha_s)$. Because σ_1 shows $\log_g(\alpha) = \log_{g_{s+c}}(\alpha_{s+c})$, they both imply $\log_g(\alpha) = \log_{g_s}(\alpha_s)$. But $\alpha_s = g_2^r$, $\alpha = g^{r_1}$ and $g_s = g^s$, hence this shows $r_2 = r_1 s$. The first method requires the knowledge of $\log_g(g_s)$, thus can only be executed by the signer. The second method requires the knowledge of $\log_g(g_c)$ and thus can only be executed by the confirmer.

5   Security Analysis

To show that the construction is secure, we assume that Schnorr signature scheme and Chaum-Petersen signature of equality of discrete logarithms are secure. Fujioka-Okamoto-Ohta Interactive bi-proof of equality is secure, sound and witness-indistinguishable. Such security proofs can be shown in random oracle model[18,10]. We now show that the construction is unforgeable, indistinguishable and signature verifications are consistent. All proofs are in the random oracle model.

5.1   Signature Unforgeability:

Definition 3. *Strong signature unforgeability property*
Except with negligible probability, there exists no probabilistic polynomial time algorithm $\mathcal{A}$ that can produce a signature σ on a specific message m, verifiable with the public key y given access to signing oracles of all the public key y^* for

all messages excluding access to y for the message m. Here given any message m^*, *the signing oracle of a public key* y^* *produces a signature* σ^* *of* m^* *verifiable with* y^*.

Intuitively, strong signature unforgeability property means that given access to signing oracles of all valid public keys for all messages excluding the desired one, it is infeasible to produce a signature σ under the desired public key, on a desired message m.

This definition is no weaker than the standard secure signature notion[13]. We consider it as the equivalent counterpart of secure against adaptive chosen ciphertext attack property of encryption schemes[19]. We say that a signature scheme is strongly unforgeable if it satisfies the strong signature unforgeability property.

Lemma 1. *In the random oracle paradigm[1], if a signature scheme is proven secure under standard secure signature definition by using the proof model of [18], the signature scheme has the strong signature unforgeability property. In other words, in the proof model of [18], standard secure signature and strong signature unforgeability property are equivalent.*

This is because the simulation of the adversary under the standard model and our model is identical. The only diffidence is, in the latter case, we allow the adversary to query not one but many signing oracles. Nonetheless, the numbers of queries must still be polynomial. The remaining of the simulation is identical to that of [18]. The simulation is repeated until the adversary gathers enough information to compute the secret key using the knowledge extractor of the proof system.

Lemma 2. *A signature* $\sigma = (\sigma_1, \sigma_2)$ *successfully passes one of the verification test only if* $\sigma_1 = \mathcal{CP}(r_1, \alpha, \alpha_{s+c}, g, g_{s+c})(m)$, $\sigma_2 = \mathcal{S}(sr_2, g, \alpha_s)(\sigma_1)$ *and* $r_1 = r_2$.

If σ is valid, σ_1 and σ_2 are formed as $\sigma_1 = \mathcal{CP}(r_1, \alpha, \alpha_{s+c}, g, g_{s+c})(m)$ and $\sigma_2 = \mathcal{S}(sr_2, g, \alpha_s)(\sigma_1)$. The remaining is to show $r_1 = r_2$. We can certainly assume that s is not 0. This is easily verified.

A signature σ is considered valid if it passes one of the two verification tests. If it is the verification test with the confirmer, the bi-proof $Bi - Proof[\log_g(g_c) \equiv \log_\alpha(\alpha_c)]$ must hold. Thus it shows that $\alpha_c = \alpha^c$ or $\alpha^c = \alpha_{s+c}/\alpha_s$. Besides as σ_1, σ_2 are sound, there exists r_1 and sr_2 such that $\alpha_{s+c} = g_{s+c}^{r_1} = g^{(s+c)r_1}$ and $\alpha_s = g^{sr_2}$. Hence we have $\alpha^c = g^{r_1c} = g^{(s+c)r_1}/g^{sr_2}$ or $g^{sr_2} = g^{sr_1}$. As $s \neq 0$, this shows $r_1 = r_2$. The case of the verification test with the signer is similar and omitted here.

Theorem 1. *In the random oracle paradigm, our undeniable confirmer signatures are unforgeable.*

Proof:

Due to lemma 2, a signature σ^+ is valid if $\sigma_1^+ = \mathcal{CP}(r_1, \alpha, \alpha_{s+c}, g, g_{s+c})(m)$, $\sigma_2^+ = \mathcal{S}(sr_2, g, \alpha_s)(\sigma_1)$ and $r_1 = r_2$. This means that if there exists a polynomial time adversary $\mathcal{A}$ that successfully create both σ_1^+ and σ_2^+, then $\mathcal{A}$ must know r_1, r_2s and the secret key s. Thus the only scenario that $\mathcal{A}$ can forge σ^+ without access to secret key s, is to obtain either σ_1^+ or σ_2^+ for the signer and create the other. W.l.o.g let us assume that $\mathcal{A}$ obtain σ_1^+. Of course, σ_1^+ comes from a signature $\sigma^* = (\sigma_1^+, \sigma_2^*)$. Due to lemma 2, this means σ_2^* and σ_2^+ are generated using the same secret key (r_2s). This means $\mathcal{A}$ knows the secret to generate σ_2^* that contradicts with the strong unforgeability property of σ_2.

Signature Indistinguishability:

Definition 4. *(Simulated signature) Given $x, g_y = g^y$ and $g_z = g^z$, a simulated signature $\sigma^* = (\sigma_1^*, \sigma_2^*)$ on a message m is computed as*

$$\sigma_1^* = \mathcal{CP}(x, X, X_{y+c}, g, g_{y+c}) \text{ and } \sigma_2^* = \mathcal{S}(z, g, y_z)(\sigma_1^*)$$

where c, g_c are confirmer's secret/public key, $X = g^x$, $X_{y+c} = g_{y+c}^x$ and $g_{y+c} = g_y g_c$.

Such signature is constructible under random oracle model. The first part of the signature σ_1^* can always be constructed as x is known. The second part of the signature σ_2^* is a Schnorr signature verifiable using the public key $g_z = g^z$. A Schnorr signature (u, v) is simulatable in random oracle model. This is done by choosing u, v at random and simulating the random oracle in such a way that it outputs u for the input $(m\|y\|g\|g^u y^v)$.

Theorem 2. *In random oracle model, if there exists a simulator $\mathcal{A}$ that can distinguish a valid signature from a simulated signature generated using the above definition in probabilistic polynomial time, there is an algorithm that solves the Diffie-Hellman decision problem in probabilistic polynomial time.*

Proof:
We prove by contradiction. We assume that there is an adversary $\mathcal{A}$ that can distinguish a valid signature σ from a simulated signature σ^* using only public information.

We use $\mathcal{A}$ to solve the Diffie-Hellman decision problem. We denote the set of all $(a, g^b, g^c | ab = c)$ as $\mathcal{D}$ and $(a, g^b, g^c | a \in_R Z_q^*)$ as $\mathcal{X}$. For given two tuples $t^* = (x_1, g^y, g^z) \in \mathcal{D}$ and $t^+ = (x_2, g^y, g^z) \in \mathcal{X}$. According to our definition of simulated signatures, $\mathcal{A}$ can generate two simulated signatures σ^* and σ^+ from t^* and t^+ respectively. Here the signer's public key is g^y.

According to lemma 2, σ^* is a valid signature; σ^+ is an invalid one. Also the advantages of $\mathcal{A}$ in distinguishing σ^* from σ^+ is negligible over distinguishing between t^* and t^+. So if $\mathcal{A}$ is able to identify the correct signature from (σ^*) and (σ^+), we can tell which of t^* or t^+ comes from $\mathcal{D}$. That completes our proof of indistinguishability.

Consistency of Signature Verification: According to lemma 2, a signature is valid only if either $\alpha_{s+c}/\alpha_s = \alpha^c$ or $\alpha_s = \alpha^s$. It is straightforward to prove that the reverse relation, namely if either $\alpha_{s+c}/\alpha_s = \alpha^c$ or $\alpha_s = \alpha^s$ holds, σ_1

is a legitimate proof of knowledge and equality and σ_2 is a valid signature, σ is a correct signature. Thus consistency of signature verification follows from the correctness and soundness of the bi-proof of knowledge.

6 Blind Undeniable Confirmer Signature and Its Application

6.1 Construction

Both Schnorr identification protocol and Chaum-Petersen interactive proof of equality are divertible[17,21]. For completeness, we formally present the blind undeniable confirmer signature generation protocol here. The protocol consists of an instance of the blind Schnorr signature and an instance of blind Chaum-Petersen signature of equality run in parallel. The detailed construction is as follows:

$\mathcal{Receiver}$ $\mathcal{Signer}$

$\rho, \tau_1, \tau_2 \in_R \mathcal{Z}_q$ $r, r_1, r_2 \in_R \mathcal{Z}_q$

$$\alpha := g^r$$
$$\alpha_s := g_s^r$$
$$\alpha_{s+c} := g_{s+c}^r$$
$$w_2 := g^{r_1}$$
$$w_1 := g^{r_2}$$
$$W_1 := g_{s+c}^{r_2}$$

$$\xleftarrow{\alpha, \alpha_s, \alpha_{s+c}, w_2, w_1, W_1}$$

$\beta := \alpha^\rho$
$\beta_s := \alpha_s^\rho$
$\beta_{s+c} := \alpha_{s+c}^\rho$
$\omega_2 := w_2^\rho g^{\tau_2}$
$\omega_1 := w_1^\rho g^{\tau_1}$
$\Omega_1 := W_1^\rho g^{\tau_1}$
$\upsilon := \mathcal{H}(m \| \beta \| \beta_s \| \beta_{s+c} \| \omega_2 \| \omega_1 \| \Omega_1)$
$u := \upsilon / \rho$

$$\xrightarrow{\quad u \quad}$$

$$v_1 := r_1 - u(r)$$
$$v_2 := r_2 - u(rs)$$

$$\xleftarrow{\quad v_1, v_2 \quad}$$

$\nu_1 := v_1 \rho + \tau_1$
$\nu_2 := v_2 \rho + \tau_2$
$\Downarrow$
$\sigma_1^+ = (\upsilon, \nu_1, \beta, \beta_{s+c}, \omega_1, \Omega_1)$
$\sigma_2^+ = (\upsilon, \nu_2, \beta_s, \omega_2)$

Here in order to obtain a blind undeniable confirmer signature $\sigma = (\sigma_1^+, \sigma_2^+)$, we divert an interactive instance of the signature generation protocol from generating $\sigma_1 = \mathcal{CP}(r, \alpha, \alpha_{s+c}, g, g_{s+c})(m)$ and $\sigma_2 = \mathcal{S}(sr, g, \alpha_s)(\sigma_1)$ to generating

$\sigma_1^+ = \mathcal{CP}(r\rho, \beta, \beta_{s+c}, g, g_{s+c})(m)$ and $\sigma_2^+ = \mathcal{S}(sr\rho, g, \beta_s)(\sigma_1^+)$ where $\beta = \alpha^\rho$, $\beta_s = \alpha_s^\rho$ and $\beta_{s+c} = \alpha_{s+c}^\rho$. Also the value ρ is known to the intermediate who is also the signature receiver in the protocol.

It is straightforward to verify that σ_1^+ and σ_2^+ constitute a valid Chaum-Petersen signature on the message m and a valid Schnorr signature on the message $(m\|\sigma_1^+)$ respectively. This implies that $\sigma^+ = (\sigma_1^+, \sigma_2^+)$ is a valid undeniable confirmer signature.

6.2   A Scalable Subscription Payment Scheme

The economy of on-line services over Internet requires fast and efficient payment mechanisms. A popular solution is micropayments where users make a small payment for every purchased on-line content. An alternative solution is subscription where each user subscripts to the service for some fixed annual fee. The user is then issued a subscription certificate that allows access to every service content. The advantage of subscription based payments over micropayments is that it removes the burden of handling a large number of small value transactions. In practice, it is very unlikely that a service provider can provide all desired services for an user. It is also inconvenient for any user to keep a number of subscription certificates. A more desired solution would be a large consortium of service providers who provide many different kinds of on-line services[1]. In order to get access to those services, each user only obtains one subscription certificate for which he has to pay some annual fees[2]. In return, the user gets full access to all of the services provided by any consortium member.

Our blind signature can be used to design a subscription system with user privacy. In this model, the subscription certificate is a blind undeniable confirmer signature issued by the consortium manager. To access to on-line services, the user shows a valid subscription certificate to the service provider who plays the role of a confirmer in the signature scheme.

Main advantages of this approach is that it removes the burden of handling large number of payments of small-amount while still provides user privacy. This is particularly desirable in Internet environment where unlike conventional subscription systems such as magazines and pay-tv, the service provider knows every content that an user has viewed. Our scheme also prevents illegal trading of subscription certificates on a commercial basis. This is because a certificate can only be verified with the help of some service providers or the consortium manager. It is possible to prevent a service provider and an user to collude by designing the certificates such that permissions to access to on-line contents provided by a service provider can only be verified by either that provider or the consortium manager.

---

[1] An alternative scenario is that a service provider providing a large number of services of which some are subcontracted to other service providers.

[2] How the fees are divided between the providers are outside the scope of this paper.

References

1. M. Bellare and P. Rogaway, "Random Oracles are practical: a paradigm for designing efficient protocols", Proc. 1st ACM Conference on Computer and Communications Security, ACM Press, 1993, pp. 62-73.
2. J. Boyar, D. Chaum, I. Damgåar and T. Pedersen, "Convertible undeniable signatures", Proc. of Crypto'90, Springer-Verlag, 1991, pp. 189-205.
3. J. Camenisch and M. Stadler, "Efficient group signature schemes for large groups", Proc. of Crypto'97, Springer-Verlag 1997, pp. 410-424.
4. D. Chaum and H. van Antwerpen, "Undeniable signatures", Proceedings of Crypto'89, Springer-Verlag, 1990, pp. 212-216.
5. D.Chaum, "Zero-knowledge undeniable signatures", Proceedings of Eurocrypt'90, Springer-Verlag, 1991, pp. 458-464.
6. D. Chaum, "Designated confirmer signatures", Proc. of Eurocrypt'94, Springer-Verlag, 1995, pp.86-91.
7. D. Chaum and T. Pedersen, "Wallet databases with observers", Proc. of Crypto'92, Springer-Verlag, 1993, pp. 89-105.
8. L. Chen, "Efficient fair exchange with verifiable confirmation of signature" in Proceedings of Asiacrypt'98, Springer-Verlag, 1998, pp. 286-298 .
9. I. Damgård and T. Pedersen, "New convertible undeniable signature schemes", Proceedings of Eurocrypt'96, Springer-Verlag, 1996, pp. 372-386.
10. A. Fujioka, T. Okamoto and K. Ohta, "Interactive bi-proof systems and undeniable signature schemes", Proc. of Eurocrypt'91, Springer-Verlag, 1992, pp. 243-256.
11. R. Gennaro, H. Krawczyk and T. Rabin, "RSA-based undeniable signatures", Proc. of Crypto'97, Springer-Verlag, 1997, pp. 132-149.
12. O. Goldreich, S. Micali and A. Wigderson, "Proofs that yield nothing but their validity and a methodology of cryptographic protocol design", Proceedings of IEEE 27th Annual Symposium on Foundations of Computer Science, pp. 174-187, 1986.
13. S. Goldwasser, S. Micali and R. Rivest, "A digital signature scheme secure against adaptive chosen-message attacks", SIAM J. Computing, 17(2):281-308, April 1988.
14. M. Michels and M. Stadler, "Efficient convertible undeniable signature schemes", Proc. of 4th Annual Workshop on Selected Areas in Cryptology(SAC'97), 1997, pp. 231-243.
15. M. Michels and M. Stadler, "Generic constructions for secure and efficient confirmer signature Schemes", Proceedings of Eurocrypt'98, Springer-Verlag, 1998, pp.405-421.
16. T. Okamoto, "Designated confirmer signatures and public key encryption", Proceedings of Crypto'94, LNCS 839, Springer-Verlag, 1994, pp. 61-74.
17. T.Okamoto and K. Ohta, "Divertible zero-knowledge interactive proofs and commutative random self-reducibility", Proceedings of Eurocrypt'89, Springer-Verlag, 1990, pp. 134-149.
18. D. Pointcheval and J. Stern, "Security proofs for signature schemes", Proceedings of Eurocrypt'96, Springer-Verlag, 1996, pp. 387-398.
19. C. Rackoff and D. Simon, "Non-interactive zero-knowledge proof of knowledge and chosen ciphertext attack", Proceedings of Crypto'91, Springer-Verlag, 1991, pp. 433-444.

20. C. Schnorr, "Efficient identification and signatures for smart cards", Proc. of Crypto'89, Springer-Verlag, 1990, pp. 239–252.
21. M. Stadler, "Cryptographic Protocols for Revocable Privacy", Ph.D Thesis, Swiss Federal Institute of Technology, Zurich, 1996.

Extended Proxy Signatures for Smart Cards

Takeshi Okamoto[1], Mitsuru Tada[1], and Eiji Okamoto[2]

[1] School of Information Science,
Japan Advanced Institute of Science and Technology,
Asahidai 1-1, Tatsunokuchi, Nomi, Ishikawa, 923-1292, Japan
{kenchan, mt}@jaist.ac.jp
[2] Center for Cryptography, Computer and Network Security(CCCNS),
University of Wisconsin, Milwaukee
Milwaukee, WI 53201, USA
okamoto@cs.uwm.edu

Abstract. Proxy signatures, which was proposed by Mambo, Usuda, and Okamoto in 1996, allow a designated person to sign on behalf of an original signer. In this paper, we present a new concept of proxy signature scheme, which can diminish the computational cost and the amount of the memory. This concept is suitable for the application of smart cards, whose processors and memories are rather limited even now. Furthermore, we propose two concrete schemes which are based on the Nyberg-Ruppel scheme and on the RSA scheme, respectively. We also consider the rigorous security of our proposed schemes by using the reduction among functions.

1   Introduction

In 1996, Mambo, Usuda, and Okamoto formulated the general idea of the proxy signature scheme [1]. This scheme allows a designated person, called a proxy signer, to sign on behalf of the original signer. In [1] they classified three types of delegation such as full delegation, partial delegation, and delegation by warrant. These types are defined as follows:

- **Full delegation**: An original signer gives just his/her own secret key to a proxy signer. So a signature created by a proxy signer is indistinguishable from that created by an original signer.
- **Partial delegation** [5,6,7]: An original signer does not give his/her own secret key, but creates a proxy's secret key from it, which will then be sent to the proxy signer. Note that the original signer's secret key should not be feasibly computed from any proxys' secret keys for the sake of its security. The proxy signer can generate a proxy signature on any message by using a proxy signer's secret key.
- **Delegation by warrant** [8,9] : This delegation is implemented by using a warrant, which guarantees that a signer is exactly a legal signer.

There are many proxy signature schemes [5 - 9] for each of these delegation types. Both of the partial delegation and the delegation by warrant have higher

M. Mambo, Y. Zheng (Eds.): ISW'99, LNCS 1729, pp. 247–258, 1999.

security than the full delegation. Compared with the delegation by warrant, the partial delegation brings faster processing speed. In 1997, Kim, Park and Won [5] revised the concept mentioned above to propose the two new types of digital proxy signatures, which are called partial delegation with warrant and threshold delegation. Zhang [7] proposed two schemes for partial delegation with nonrepudiable property, which means it is possible to decide who is the actual signer of a proxy signature. Therefore, in these schemes only the proxy signer can create a valid signature scheme for the original signer. In 1998, Lee, Hwang and Wang [6] pointed out that Zhang's second scheme [5] does not have this property and showed that a dishonest proxy signer can cheat to get an original signer' signature. They also modified the scheme to prevent such attacks and included a feature that the original signer can limit the delegation time for a certain period.

In this paper, we seek another goal to study. That is, our study does not include the topics mentioned above. To utilize the property with the proxy signatures for partial delegation, we apply it to an off-line e-cash system using smart card, which has the following features:

- **Smart card**: A card of this type contains an embedded microprocessor chip capable of both storing and processing. This property allows us to make purchases from a credit account, debit account, or stored value on the card.

We show an example of how to apply a proxy signature to an off-line e-cash system for smart card. This e-cash system consists of three parties, namely an original signer *bank*, a proxy signer *user*, and a verifier *shop*. At first, the bank makes a smart card for the user. In this card, the bank puts a message of conditions such as a situation of the user, the limit of money which s/he can spend, the period for which the card is available, and so on. The bank digitally signs the message, puts the signature in the card, and hands it to the user. When the user buys something at the shop by using a smart card, s/he signs the message in which the price is described. The clerk of the shop confirms the validity of both the bank's and the user's signatures. If the verification holds, the clerk hands the shop's merchandise to the user, and deposits the received e-cash afterward. To implement this system, we must pass the severe conditions, because the processors and memories of a current smart card are now rather limited. Our goal in this paper is to propose schemes which can diminish the computational cost and the amount of the memory.

This paper is organized as follows. In Section 2, we will review the previous work on partial delegation. In Section 3, we will indicate the basic idea of the proposed schemes and classify the problems for them. In Sections 4 and 5, we will propose two new proxy signature schemes and show that our proposed schemes are provably secure by using the concept of reductions among functions. Finally, concluding remarks will be given in Section 6.

2   Previous Work

We review the basic protocol for partial delegation in [1].

Original Parameter Generation Phase

- **Step 1.** (Original generation) Let p be a prime, q be a prime such that $q|p - 1$, and $g \in Z_p^*$ be an order-q element. An original signer generates a random number $x \in Z_q$ which s/he keeps secret, computes $y = g^x \bmod p$, and makes (p, q, g, y) public.

Proxy Parameter Generation Phase

- **Step 2.** (Proxy generation) An original signer generates a random number $k \in Z_q$ and computes:

$$K = g^k \qquad \bmod p;$$
$$\sigma = x + kK \bmod q.$$

- **Step 3.** (Proxy delivery) The original signer publishes K and delivers σ to a proxy signer through a secure channel.
- **Step 4.** (Proxy verification) The proxy signer checks whether the following verification holds:

$$g^\sigma \stackrel{?}{=} yK^K \bmod p.$$

If it does, then s/he accepts those as valid proxy parameters. Otherwise s/he rejects those.

Signature and Verification Phase

- **Step 5.** (Signing by the proxy signer) To sign a message m_p on behalf of the original signer, s/he can use the proxy signature key σ as an alternative of x and generates a signature $Sign_\sigma(m_p)$ by executing a conventional digital signing operation. Then the created proxy signature is $(m_p, Sign_\sigma(m_p), K)$.
- **Step 6.** (Verification of the proxy signature) The verification of the proxy signature is carried out by the same checking operation as in the original signature scheme except for replacing with $y' = yK^K \bmod p$.

3   Basic Idea and Classification of Problems

We describe the basic idea which is used by our proposed schemes.

Definition 3.1 [Extended Proxy Signature Scheme] This scheme is classified in partial delegation of proxy signature scheme. Therefore, an original signer has a secret key x, creates a new secret key σ from x, and sends σ to a proxy signer in a secure way. Note that this scheme has a property that it is hard for the proxy signer to compute s from σ with any feasible computational cost. When the proxy signer signs a message m_p, s/he generates a proxy signature $Sign_\sigma(m_p)$ by using σ and sends it to a verifier. The verifier checks the validity of this signature. The verifier then accepts it if the verifier recovers the message which the original signer has already signed. Otherwise s/he rejects it. For the sake of its security, this scheme must have the condition that the verifier can not recovered any message which not the original signer, but the proxy signer has freely signed.

In the verification step, the verifier can confirm the validity of both a proxy signer's message and an original signer's one. Note that only one check is sufficient for the validity, and if the signature is valid, the verifier can recover a message signed by the original signer. This means that the proxy signer have only to sends the proxy signer's message and signature, and does not have to send the original signer's message. This property is profitable because this scheme can diminish the computational cost and the amount of the memory.

Next, we give the classification of problems with respect to the security of extended proxy signature scheme as follows.

Problem 1 [Impersonation of Original Signer by Dishonest Proxy Signer] This is a problem that a dishonest proxy signer breaks a proposed scheme by computing an original signer's secret key. Therefore, the proxy signer can impersonate not only the original signer but any other proxy signers.

Problem 2 [Impersonation of Original Signer by Malicious Attacker] This is a problem that a malicious attacker who is not a designated proxy signer, breaks a proposed scheme by computing an original signer's secret key. Therefore, the attacker can impersonate not only the original signer but any proxy signers.

Problem 3 [Impersonation of Another Proxy Signer by Dishonest Proxy Signer] This is a problem that a dishonest proxy signer breaks a proposed scheme by computing any other proxy signers' secret keys. Therefore, the proxy signer can impersonate any other proxy signers.

Problem 4 [Impersonation of Proxy Signer by Malicious Attacker] This is a problem that a malicious attacker which is not a designated proxy signer, breaks the proposed scheme by computing any proxy signers' secret keys. Therefore, the attacker can impersonate any proxy signers.

Our proposed schemes must not be broken on the problems mentioned above. Therefore, when we propose a new scheme, we must prove the security against these problems.

4 Nyberg-Ruppel-Based Proxy Signature Scheme

We describe an extended proxy signature scheme, which is based on the Nyberg-Ruppel message recovery signature scheme [11] (see Appendix). This scheme includes the property shown in Section 3.

4.1 Proposed Scheme 1

Original Parameter Generation Phase

- **Step 1.** (Original generation) Let p be a prime, q be a prime such that $q|p-1$, and $g \in Z_p^*$ be an order-q element. An original signer generates a random number $x \in Z_q$ which s/he keeps secret, computes $y = g^x \bmod p$, puts (p, q, g, y) in a smart card, and makes them public.

Proxy Parameter Generation Phase

- **Step 2.** (Proxy generation) The original signer generates a random number $\kappa \in Z_q$ and computes:

$$\rho = (ID_p \| m_w) g^{-\kappa} \bmod p;$$
$$\sigma = -xh(\rho, I_p) + \kappa \bmod q,$$

where ID_p, m_w, and I_p mean a proxy signer's identity information, a warrant message such as a situation of the user, the limit of money which s/he can spend, and so on, and additional proxy signer's information, respectively. Moreover, we also define $\|$ is the operation of concatenation, and h is a collision-resident hash function mapping $h : \{0,1\}^* \mapsto \{0,1\}^{|t|-1}$ with the security parameter t.
- **Step 3.** (Proxy delivery) The original signer puts (ρ, ID_p, I_p) which are made public and σ which is kept secret in a smart card, and sends this card to the proxy signer.
- **Step 4.** (Proxy verification) The proxy signer checks whether the following verification holds:

$$(ID_p \| m_w) \stackrel{?}{=} g^\sigma y^{h(\rho, I_p)} \rho \bmod p.$$

If it does, then the proxy signer accepts those as valid proxy parameters. Otherwise s/he rejects those.

Signature and Verification Phase

- **Step 5.** (Signing by the proxy signer) To sign a message m_p on behalf of the original signer, the proxy signer generates a random number $k \in Z_q$, and computes:

$$r = g^{-k} \qquad \bmod p;$$
$$s = \sigma + kh(r, m_p) \bmod q,$$

and sends them to a verifier. Then the created proxy signature is (m_p, (r, s), (ρ, I_p)). We define $v = y^{h(\rho, I_p)} \rho \bmod p$, which is regarded as a proxy signer's new public information.

- **Step 6.** (Verification of the proxy signature) The verifier checks whether the following verification holds:

$$(ID_p||m_w) \overset{?}{=} g^s v r^{h(r,m_p)} \bmod p,$$

If it does, then the verifier accepts it as a valid proxy signature. Otherwise s/he rejects it.

In Step 6, if the proxy signature is valid, then the verification holds since

$$\begin{aligned}
g^s v r^{h(r,m_p)} \bmod p &= g^{\sigma + kh(r,m_p)} y^{h(\rho,I_p)} \rho g^{-kh(r,m_p)} \\
&= g^{-xh(\rho,I_p)+\kappa+kh(r,m_p)} g^{xh(\rho,I_p)} (ID_p||m_w) g^{-\kappa} g^{-kh(r,m_p)} \\
&= (ID_p||m_w) \bmod p.
\end{aligned}$$

4.2   Security Considerations (Proposed Scheme 1)

We discuss the security of our proposed schemes by using reductions among functions. At first, polynomial-time many-one reducibility is defined as follows.

Definition 4.1 [Polynomial-Time Many-One Reducibility] For functions F and G, if there exists a pair of polynomial-time computable functions (Z_1, Z_2) such that $F(x) = Z_1(G(Z_2(x)))$, then we say "F reduces to G with respect to the polynomial-time many-one reducibility", and write as $F \leq_m^{FP} G$. If the converse reduction also holds, we say "F is equivalent to G with respect to the polynomial-time many-one reducibility", and write as $F \equiv_m^{FP} G$.

We define a functions to solve the discrete logarithm problem. This way is similar to the definitions for some discrete logarithm cryptosystems shown in [18].

Definition 4.2 [Discrete Logarithm Problem] $\text{DLP}(p, g, y')$ is a function that on input $p \in N_{prime}$, $g \in Z_p^*$, $y' \in Z_p^*$, outputs $x' \in Z_{p-1}^*$ such that $y' = g^{x'} \bmod p$, if such an x' exists.

We describe other functions, which are defined to solve the problems shown in Section 3, and conclude the results on difficulties of breaking those functions.

Definition 4.3 [Problem 1 on Proposed Scheme 1] Ps1-P1 $(p, q, g, y, \rho, \sigma, ID_p, m_w, I_p)$ is a function that on input $p \in N_{prime}$, $q|p-1$, $q \in N_{>1}$, $g \in Z_p^*$, $y \in Z_p^*$, $\rho \in Z_p^*$, $\sigma \in Z_q^*$, $(ID_p||m_w) \in Z_p$, $I_p \in N$, outputs $x \in Z_q^*$ such that $y = g^x \bmod p$, $\rho = (ID_p||m_w) g^{-\kappa} \bmod p$, $\sigma = -xh(\rho, I_p) + \kappa \bmod q$, $\kappa \in Z_q$, $h(\rho, I_p) \in Z_{2^t}$, if such an x exists.

Theorem 4.4 $\text{Ps1-P1} \equiv_m^{FP} \text{DLP}$.

Proof.

1. $\text{Ps1-P1} \leq_m^{FP} \text{DLP}$:

$$\text{Ps1-P1}(p, q, g, y, \rho, \sigma, ID_p, m_w, I_p) = \text{DLP}(p, g, y).$$

2. DLP $\leq_m^{FP}$ Ps1-P1:

$$\text{DLP}(p, g, y') = \text{Ps1-P1}(p, p - 1, g, y', 1, 1, 0, 1, 1).$$

∎

Definition 4.5 [Problem 2 on Proposed Scheme 1] Ps1-P2$(p,\ q,\ g,\ y,\ \rho,$ $r,\ s,\ ID_p,\ m_w,\ m_p,\ I_p)$ is a function that on input $p \in N_{prime}$, $q | p - 1$, $q \in$ $N_{>1}$, $g \in Z_p^*$, $y \in Z_p^*$, $\rho \in Z_p^*$, $r \in Z_p^*$, $s \in Z_q^*$, $(ID_p || m_w) \in Z_p$, $m_p \in N$, $I_p \in N$, outputs $x \in Z_q^*$ such that $y = g^x \bmod p$, $\rho = (ID_p || m_w) g^{-\kappa} \bmod p$, $\kappa \in Z_q$, $r = g^{-k} \bmod p$, $k \in Z_q$, $s = \sigma + kh(r, m_p) \bmod q$, $h(r, m_p) \in Z_{2^t}$, $\sigma = xh(\rho, I_p) + \kappa \bmod q$, $h(\rho, I_p) \in Z_{2^t}$, if such an x exists.

Theorem 4.6 Ps1-P2 $\equiv_m^{FP}$ DLP.

Proof.

1. Ps1-P2 $\leq_m^{FP}$ DLP:

$$\text{Ps1-P2}(p, q, g, y, \rho, r, s, ID_p, m_w, m_p, I_p) = \text{DLP}(p, g, y).$$

2. DLP $\leq_m^{FP}$ Ps1-P2:

$$\text{DLP}(p, g, y') = \text{Ps1-P2}(p, p - 1, g, y', 1, 1, 1, 0, 1, 1, 1).$$

∎

Definition 4.7 [Problem 3 on Proposed Scheme 1] Ps1-P3$(p,\ q,\ g,\ y,\ \rho,\ \sigma,$ $ID_p,\ m_w,\ I_p,\ \rho_i,\ ID_{pi},\ m_{wi},\ I_{pi})$ is a function that on input $p \in N_{prime}$, $q | p - 1$, $q \in N_{>1}$, $g \in Z_p^*$, $y \in Z_p^*$, $\rho \in Z_p^*$, $\sigma \in Z_q^*$, $(ID_p || m_w) \in Z_p$, $I_p \in N$, $\rho_i \in Z_p^*$, $(ID_{pi} || m_{wi}) \in Z_p$, $I_{pi} \in N$, where i means a user's name such as $i = A, B, C, \ldots$, outputs $\sigma_i \in Z_q^*$ such that $y = g^x \bmod p$, $x \in Z_q$, $\rho = (ID_p || m_w) g^{-\kappa} \bmod p$, $\kappa \in Z_q$, $\sigma = -xh(\rho, I_p) + \kappa \bmod q$, $h(\rho, I_p) \in Z_{2^t}$, $\rho_i = (ID_{pi} || m_{wi}) g^{-\kappa_i} \bmod p$, $\kappa_i \in Z_q$, $\sigma_i = -xh(\rho_i, I_{pi}) + \kappa_i \bmod q$, $h(\rho_i, I_{pi}) \in Z_{2^t}$, if such a σ_i exists.

Theorem 4.8 Ps1-P3 $\equiv_m^{FP}$ DLP.

Proof.

1. Ps1-P3 $\leq_m^{FP}$ DLP:

$$\text{Ps1-P3}(p, q, g, y, \rho, \sigma, ID_p, m_w, I_p, \rho_i, ID_{pi}, m_{wi}, I_{pi}) = -\text{DLP}(p, g, y)\, h(\rho_i, I_{pi})$$

2. DLP $\leq_m^{FP}$ Ps1-P3:

$$\text{DLP}(p, g, y') = -\text{Ps1-P3}(p, p - 1, g, y', 1, 1, 0, 1, 1, 1, 0, 1, 1)\, h(1, 1).$$

∎

Definition 4.9 [Problem 4 on Proposed Scheme 1] $\text{Ps1-P4}(p, q, g, y, \rho, r,$
$s, ID_p, m_w, I_p, \rho_i, ID_{pi}, m_{wi}, I_{pi})$ is a function that on input $p \in N_{prime}, q|p-1,$
$q \in N_{>1}, g \in Z_p^*, y \in Z_p^*, \rho \in Z_p^*, r \in Z_p^*, s \in Z_q^*, (ID_p||m_w) \in Z_p, I_p \in N,$
$\rho_i \in Z_p^*, (ID_{pi}||m_{wi}) \in Z_p, I_{pi} \in N$, outputs $\sigma_i \in Z_q^*$ such that $y = g^x \bmod p,$
$x \in Z_q, \rho = (ID_p||m_w)g^{-\kappa} \bmod p, \kappa \in Z_q, r = g^{-k} \bmod p, k \in Z_q, s = \sigma +$
$kh(r, m_p) \bmod q, \sigma = -xh(\rho, I_p) + \kappa \bmod q, h(r, m_p) \in Z_{2^t}, h(\rho, I_p) \in Z_{2^t}, \rho_i =$
$(ID_{pi}||m_{wi})g^{-\kappa_i} \bmod p, \kappa_i \in Z_q, \sigma_i = -xh(\rho_i, I_{pi}) + \kappa_i \bmod q, h(\rho_i, I_{pi}) \in Z_{2^t},$
if such a σ_i exists.

Theorem 4.10 $\text{Ps1-P4} \equiv_m^{FP} \text{DLP}.$

Proof.

1. $\text{Ps1-P4} \leq_m^{FP} \text{DLP}$:

   $\text{Ps1-P4}(p, q, g, y, \rho, r, s, ID_p, m_w, I_p, \rho_i, ID_{pi}, m_{wi}, I_{pi})$

   $$= - \text{DLP}(p, \; g, \; y) \; h(\rho_i, \; I_{pi}).$$

2. $\text{DLP} \leq_m^{FP} \text{Ps1-P4}$:

   $$\text{DLP}(p, g, y') = - \text{Ps1-P4}(p, p-1, g, y', 1, 1, 1, 0, 1, 1, 1, 0, 1, 1) \; h(1, 1).$$

   ∎

5 RSA-Based Proxy Signature Scheme

In the same way as Section 4, we describe another extended proxy signature
scheme, which is based on the RSA public-key cryptosystem [11].

5.1 Proposed Scheme 2

Original Parameter Generation Phase

- **Step 1.** (Original generation) An original signer picks up two primes p and
  q, puts (n, g, e) in a smart card, and makes them public, where $n = pq$, g is
  a generator of both Z_p^* and Z_q^*, and $e \in Z_{\lambda(n)}^*$. The Carmichael function of
  n is given by $\lambda(n) = \text{lcm}\,(p-1, q-1)$. Let $d \in Z_{\lambda(n)}^*$ be the secret key of
  the original signer satisfying $ed = 1 \bmod \lambda(n)$.

Proxy Parameter Generation Phase

- **Step 2.** (Proxy generation) The original signer generates

  $$\bar{\sigma} = (h(I_p)^{-1}(ID_p||m_w))^d \bmod n.$$

- **Step 3.** (Proxy delivery) The original signer puts (ID_p, I_p) which are made public and σ which is kept secret in smart card, and sends this card to the proxy signer.
- **Step 4.** (Proxy verification) The proxy signer checks whether the following verification holds:

$$(ID_p||m_w) \overset{?}{=} h(I_p) \, \bar{\sigma}^{\,e} \bmod n.$$

If does, then the proxy signer accepts those as valid proxy parameters. Otherwise s/he rejects those.

Signature and Verification Phase

- **Step 5.** (Signing by the proxy signer) To sign a message m_p on behalf of the original signer, the proxy signer generates a random number $\bar{k} \in Z_n$, and computes:

$$\bar{r} = g^{\bar{k}h(m_p)}\bar{\sigma} \bmod n;$$
$$\bar{s} = g^{-e\bar{k}} \qquad \bmod n,$$

and sends them to a verifier. Then the created proxy signature is $(m_p, (\bar{r}, \bar{s}), I_p)$.
- **Step 6.** (Verification of the proxy signature) The verifier checks whether the following verification holds:

$$(ID_p||m_w) \overset{?}{=} h(I_p) \, \bar{r}^{\,e} \, \bar{s}^{\,h(m_p)} \bmod n.$$

If it does, then the verifier accepts it as a valid proxy signature. Otherwise s/he rejects it.

In Step 6, if the proxy signature is valid, then the verification holds since

$$\begin{aligned}
h(I_p) \, \bar{r}^{\,e} \, \bar{s}^{\,h(m_p)} \bmod n &= h(I_p) \, (g^{\bar{k}h(m_p)}\bar{\sigma})^e (g^{-e\bar{k}})^{h(m_p)} \\
&= h(I_p) \, g^{e\bar{k}h(m_p)}(h(I_p)^{-1}(ID_p||m_w))^{ed}g^{-e\bar{k}h(m_p)} \\
&= (ID_p||m_w) \quad \bmod n.
\end{aligned}$$

5.2 Security Considerations (Proposed Scheme 2)

We discuss the security on the proposed scheme 2. How to discuss is similar to that in Section 4.2. At first, we define the function to break the RSA scheme [13].

Definition 5.1 [RSA Public-Key Cryptosystem] $RSA(n, e, \bar{y})$ is a function that on input $n \in N_{>1}$, $e \in Z^*_{\lambda(n)}$, $\bar{y} \in Z^*_n$, outputs $\bar{x} \in Z^*_n$ such that $\bar{y} = \bar{x}^e \bmod n$, if such an $\bar{x}$ exists.

We define other functions to solve the problems given in Section 3, and indicate the results on difficulties to compute those functions. We assume that the base g is just in Z^*_n. This assumption includes the case of g being a primitive root modulo both p and q. Note that functions to solve the problems 1 and 3 are the same in this case. This condition is similar to that of the problems 2 and 4. Therefore, we have only to consider the security on two functions.

Definition 5.2 [Problems 1 and 3 on Proposed Scheme 2] Ps2-P13$(n, e, g, \bar{\sigma}, ID_p, m_w, h(I_p), ID_{pi}, m_{wi}, h(I_{pi}))$ is a function that on input $n \in N_{>1}$, $e \in Z^*_{\lambda(n)}$, $g \in Z^*_n$, $\bar{\sigma} \in Z^*_n$, $(ID_p \| m_w) \in Z_n$, $h(I_p) \in Z_{2^t}$, $(ID_{pi} \| m_{wi}) \in Z_n$, $h(I_{pi}) \in Z_{2^t}$ outputs $\bar{\sigma}_i \in Z^*_n$ such that $\bar{\sigma} = (h(I_p)^{-1}(ID_p \| m_w))^d \bmod n$, $\bar{\sigma}_i = (h(I_{pi})^{-1}(ID_{pi} \| m_{wi}))^d \bmod n$, if such a $\bar{\sigma}_i$ exists.

Theorem 5.3 Ps2-P13 $\equiv^{FP}_m$ RSA.

Proof.

1. Ps2-P13 $\leq^{FP}_m$ RSA:

   Ps2-P3$(n, e, g, \bar{\sigma}, ID_p, m_w, h(I_p), ID_{pi}, m_{wi}, h(I_{pi}))$

   $$= \text{RSA}(n, e, h(I_{pi})^{-1}(ID_{pi} \| m_{wi})).$$

2. RSA $\leq^{FP}_m$ Ps2-P13:

   $$\text{RSA}(n, e, \bar{y}) = \text{Ps2-P3}(n, e, 1, 1, 0, 1, 1, 0, 1, \bar{y}^{-1}).$$

∎

Definition 5.4 [Problems 2 and 4 on Proposed Scheme 2] Ps2-P24$(n, e, g, \bar{r}, \bar{s}, ID_p, m_w, h(I_p), ID_{pi}, m_{wi}, h(I_{pi}))$ is a function that on input $n \in N_{>1}$, $e \in Z^*_{\lambda(n)}$, $g \in Z^*_n$, $\bar{r} \in Z^*_n$, $\bar{s} \in Z^*_n$, $(ID_p \| m_w) \in Z_n$, $h(I_p) \in Z_{2^t}$, $(ID_{pi} \| m_{wi}) \in Z_n$, $h(I_{pi}) \in Z_{2^t}$, outputs $\bar{\sigma}_i \in Z^*_n$ such that $\bar{r} = g^{\bar{k}h(m_p)}\bar{\sigma} \bmod n$, $\bar{s} = g^{-e\bar{k}} \bmod n$, $h(m_p) \in Z_{2^t}$, $\bar{k} \in Z_n$, $\bar{\sigma} = (h(I_p)^{-1}(ID_p \| m_w))^d \bmod n$, $\bar{\sigma}_i = (h(I_{pi})^{-1}(ID_{pi} \| m_{wi}))^d \bmod n$, if such an $\bar{\sigma}_i$ exists.

Theorem 5.5 Ps2-P24 $\equiv^{FP}_m$ RSA.

Proof.

1. Ps2-P24 $\leq^{FP}_m$ RSA:

   Ps2-P4$(n, e, g, \bar{r}, \bar{s}, ID_p, m_w, h(I_p), ID_{pi}, m_{wi}, h(I_{pi}))$

   $$= \text{RSA}(n, e, h(I_{pi})^{-1}(ID_{pi} \| m_{wi})).$$

2. RSA $\leq^{FP}_m$ Ps2-P24:

   $$\text{RSA}(n, e, \bar{y}) = \text{Ps2-P4}(n, e, 1, 1, 1, 0, 1, 1, 0, 1, \bar{y}^{-1}).$$

∎

6 Concluding Remarks

We have studied a new kind of a proxy signature scheme and proposed two concrete schemes, which are based on the Nyberg-Ruppel scheme and on the RSA scheme, respectively. These schemes can diminish the computational cost and the amount of the memory. Therefore, we have shown that this schemes are suitable for the usage of smart cards whose processors and memories are rather limited even now. As a consequence, we can conclude that the breaking of our proposed schemes are equivalent to breaking the discrete logarithm problem on the scheme 1 and the RSA scheme on the scheme 2, in the sense of polynomial-time reducibility. As future work, we can consider applications of the proposed scheme with respect to other key management scheme.

Acknowledgments

The authors are grateful to Prof. Atsuko Miyaji and Dr. Hisao Sakazaki for their helpful suggestions and encouragement. We appreciate the invaluable comments from anonymous referees.

References

1. M.Mambo, K.Usuda and E.Okamoto, "Proxy signatures: Delegation of the power to sign messages", *IEICE Trans. Fundamentals*, vol.E79-A, no.9, pp.1338-1354, 1996.
2. M.Mambo, K.Usuda and E.Okamoto, "Proxy signatures for delegating signing operation", *Proceedings of 3rd ACM Conference on Computer and Communications Security*, ACM press, pp.48-57, 1996.
3. M.Mambo and E.Okamoto, "Proxy cryptosystems: Delegation of the Power to Decrypt Ciphertexts", *IEICE Transactions on Fundamentals*, Vol.E80-A, no.1, pp.54-63, 1997.
4. K.Usuda, M.Mambo, T.Uyematsu and E.Okamoto, "Proposal of an automatic signature scheme using a compiler", *IEICE Trans. Fundamentals*, Vol.E79-A, no.1 pp.94-101, 1996.
5. S.Kim, S.Park and D.Won, "Proxy Signatures, Revisited", *Proceedings of ICICS'97*, LNCS 1334, Springer-Verlag, pp.223-232, 1997.
6. N.Y.Lee, T.Hwang and C.H.Wang, "On Zhang's nonrepudiable proxy signature schemes", *Third Australasian Conference, ACISP'98*, LNCS 1438, Springer-Verlag pp.415-422, 1998.
7. K.Zhang, "Threshold proxy signature schemes", *Proceedings of 1997 Information Security Workshop*, LNCS 1396, Springer-Verlag, pp.191-197, 1997.
8. B.C.Neuman, "Proxy-based authorization and accounting for distributed systems", *Proceedings of 13th International Conference on Distributed Systems*, LNCS 1438, Springer-Verlag pp.283-291, 1993.
9. V.Varadharajan, P.Allen and S.Black, "An analysis of the proxy problem in distributed systems", *Proceedings of 1991 IEEE Computer Society Symposium on Research in Security and Privacy*, LNCS 1438, Springer-Verlag pp.255-275, 1991.

10. T.ElGamal, "A public-key cryptosystem and a signature scheme based on discrete logarithms", *IEEE Trans. Inf. Theory*, vol.IT-31, no.4, pp.469-472, 1985.

11. K.Nyberg and R.A.Ruppel, "A new signature scheme based on the DSA giving message recovery", *Proceedings of 1st ACM Conference on Computer and Communications Security*, ACM press, 1993.

12. E.Okamoto and K.Tanaka, "Key distribution system based on identification information", *IEEE J. Selected Areas in Communications*, Vol.7, pp.481-485, 1989.

13. R.Rivest, A.Shamir and L.Adleman, "A method for obtaining digital signatures and public-key cryptosystem", *Communications of ACM*, vol.21, pp.120-126, 1978.

14. A.Shamir, "Identity-based cryptosystems and signature schemes", *Advances in Cryptology - CRYPTO'84*, LNCS 196, Springer-Verlag, pp.47-53, 1985.

15. C.P.Schnorr, "Efficient signature generation by smart cards", *Journal of Cryptology*, vol. 4, no. 3, pp.161-174, 1978.

16. E.Bach, "Discrete logarithms and factoring", *Technical Report UCB/CSD 84/186*, University of California, Computer Science Division (EECS), 1994.

17. M.Mambo and H.Shizuya, "A Note on the complexity of breaking Okamoto-Tanaka ID-based key exchange scheme", *Proceedings of Public Key Cryptography'98*, LNCS 1431, Springer-Verlag, pp.258-262, 1998.

18. K.Sakurai and H.Shizuya, "A structural comparison of the computational difficulty of breaking discrete log cryptosystems", *Journal of Cryptology*, Vol. 11, no 1, pp.29-43, 1998.

19. H.Woll, "Reductions among number theoretic problems", *Information and Computation*, Vol.72, pp.167-179, 1987.

Appendix. Nyberg-Ruppel Signature Scheme

This digital signature scheme has the feature that the signed message can be recovered from the signature itself. A signer choose a prime p ,a prime factor such that $q|p - 1$ and an element q of order g. The signer generates a random number $\tilde{x} \in Z_p^*$ which s/he keeps secret, and computes $\tilde{y} = g^{\tilde{x}} \bmod p$ to be public. When the signer sign a message m, s/he generates a random number $\tilde{k} \in Z_q$ and computes:

$$\tilde{r} = mg^{-\tilde{k}} \quad \bmod p,$$
$$\tilde{s} = -\tilde{x}\tilde{r} + \tilde{k} \bmod q.$$

The verifier checks whether the following verification holds:

$$m \overset{?}{=} g^{\tilde{s}}\tilde{y}^{\tilde{r}}\tilde{r} \bmod p.$$

If it is true, the verifier accepts it as a valid signature. Otherwise s/he rejects it.

A New Digital Signature Scheme on ID-Based Key-Sharing Infrastructures*

Tsuyoshi Nishioka[1], Goichiro Hanaoka[2]**, and Hideki Imai[2]

[1] Information Technology R&D Center, Mitsubishi Electric Corporation
5-1-1 Ofuna, Kamakura, 247-8501, JAPAN
Phone: +81-467-41-2181 & Fax: +81-467-41-2185
nishioka@iss.isl.melco.co.jp
[2] The 3rd Department, Institute of Industrial Science, the University of Tokyo
7-22-1 Roppongi, Minato-ku, Tokyo 106-8558, JAPAN
Phone & Fax: +81-3-3402-7365
hanaoka@imailab.iis.u-tokyo.ac.jp
imai@iis.u-tokyo.ac.jp

Abstract. Recently, many researchers have been working on ID-based key sharing schemes. The Key Predistiribution Systems (KPS) are a large class of such key sharing schemes. The remarkable property of KPS is that in order to share a key, a participant should only input its partner's identifier to its own secret-algorithm. In this paper, we propose a new signature scheme on the KPS infrastructure. Namely, it is shown that if an ID-based key sharing system which belongs to KPS is provided, a digital signature scheme can easily be realized on top of it. Moreover, this signature scheme is secure if the discrete logarithm problem is hard to solve. Although there already exists a digital signature scheme based on KPS, it has two flaws that its verifier is designated and that tamper resitstant module is needed. Our proposal solves these problems. Any entity can authenticate the signature in the new signature scheme which is based on inherence of key generator itself instead of common key. Moreover, tamper resistant module is not necessarily needed. We introduce the new concept of "one-way homomorphism" in order to realize our proposal.

1 Introduction

The concept of identity-based cryptosystem has first been proposed by Shamir in 1984[1]. Identity-based cryptosystem has been considered as third cryptosystem following common-key cryptosystem and public-key cryptosystem and many researches have been challenged by its realization[2]-[18]. This means that key management has been one of the central and most important problems. The problem is naturally important in the common-key cryptosystem and also cannot be disregarded in the public-key cryptosystem due that public-key is public

---

* A part of this work was performed in part of Research for the Future Program (RFTF) supported by Japan Society for the Promotion of Science (JSPS) under contact no. JSPS-RETF 96P00604.
** A Research Fellow of JSPS

M. Mambo, Y. Zheng (Eds.): ISW'99, LNCS 1729, pp. 259–270, 1999.

information should be authenticated public information. On the other hand, identifier, in short ID, is the most popular authenticated public information. Cryptographic infrastructure in a large-scale open network is desired to be constructed based on such authenticated public information.

Among ID-based key sharing systems, Blom's scheme[2], which has been generalized by Matsumoto and Imai[3], has very good properties in terms of computational complexity and non-interactivity. Many useful schemes based on Blom's scheme have been proposed[3,4,5,6,7,8], and these are called Key Predistribution Systems (KPS).

In a KPS, no previous communication is required and its key-distribution procedure consists of simple calculations. Furthermore in order to share the key, a participant should only input its partner's identifier to its secret KPS-algorithm. Blundo et al.[7,9], Kurosawa et al.[10] showed a lower bound of memory size of users' secret algorithms and developed KPS for a conference-key distribution.

In this paper, we propose a new signature scheme on the KPS infrastructure. Namely, it is shown that if an ID-based key sharing system which belongs to KPS is provided, a digital signature scheme can easily be realized on top of it. Moreover, this signature scheme is secure if the discrete logarithm problem is hard to solve. Although there already exists a digital signature scheme based on KPS, it has the following two flaws: its verifier is designated and a tamper resitstant module(TRM) is needed[11,19].

In the new scheme, the verifier is undesignated without losing systematic robustness. To realize it, we introduce the new concept of one-way homomorphism. The scheme is, then, based on the inherence of secret-algorithm itself instead of specific common key using the one-way homomorphism. As the results, we obtained a very sophisticated digital signature scheme.

It makes use of one-way homomorphism to provide that one-way transformed key generator is opened to the public as digital signature. We realize the scheme using one-way homomorphism on the discrete logarithm problem and the linear scheme for the KPS.

We give a brief review of KPS, its conventional signature scheme, and its basic realization in section 2. New signature scheme is proposed in section 3, and in section 4, the examples are given. We discuss the security and performance in section 5. Finally, the summary of results is given in section 6.

2   Brief Review of KPS

2.1   Property of KPS

KPS (Key Predistribution System)[3] consists of two kind of entities: one or several centers and many users. KPS-center keeps in secret the "center-algorithm,"

$$G(\cdot, \cdot), \tag{1}$$

which is bi-symmetric algorithm. The center-algorithm outputs the user's "secret-algorithm" by inputting the user's ID:

$$X_A(\cdot) = G(ID_A, \cdot), \tag{2}$$

where ID_A is the identifier of a user A and it is desirable that ID_A is a publicly authenticated symbol, for example, name, e-mail address, and so on.

Each secret-algorithm is (pre-)distributed to the user confidentially. The secret-algorithm outputs a common key between its owner and his partner simply by inputting his partner's ID:

$$k_{AB} = X_A(ID_B), \tag{3}$$
$$= X_B(ID_A), \tag{4}$$

where ID_B is identifier to the partner B and k_{AB} is common key between them.

Blundo et al.[9] showed the lower bound of required memory size for users. Suppose that for each $P \subseteq \{u_1, u_2, \cdots, u_N\}$ such that $|P| = t$, there is a key k_P associated with P. And, each user $u_i \in P$ can compute k_P, and if $F \subseteq \{u_1, u_2, \cdots, u_N\}, |F| \leq \omega$ and $|F \cap P| = 0$, then F cannot obtain any information about k_P. Then, the lower bound of the amount of users' secret algorithm X_{u_i} is estimated as follows:

$$\log_2 |X_{u_i}| \geq \binom{t + \omega - 1}{t - 1} H(K), \tag{5}$$

where $k_P \in K$ for any P. For $t = 2$, Eq. 5 becomes $\log_2 |X_{u_i}| \geq (\omega+1)H(K)$. The linear scheme[3], Blom[2], Blundo et al.[7] and some other schemes are known as optimal schemes to achieve the lower bound. In order to achieve perfect security, $\omega + t$ must be equal to the number of entities in the whole network. Hence, for $t = 2$ if $\omega = N - 2$ is selected, the whole system becomes unconditionally secure by using the optimal schemes.

2.2 Conventional Digital Signature Scheme

By using the KPS technology, a simple digital signature scheme can be provided. Although we can realize this signature scheme easily, it has two serious problems:1)The signature designates only the verifier and any other entities cannot authenticate the signature, 2)This scheme requires TRM. The main contribution of this article is to solve these two problems. Moreover, our scheme is secure if the discrete logarithm problem is hard to solve.

Here, we give the brief review of the conventional digital signature scheme based on KPS. It prepares an additional secret-algorithm for each entities. Each entity has, then, two secret-algorithms: the secret-algorithm for general purpose X_A and the other secret-algorithm for authentication:

$$X_{A_V}(\cdot) = G(ID_{A_V}, \cdot), \tag{6}$$

where ID_{A_V} is identifier for authentication and usually given by the ID transformation f_V:

$$ID_{A_V} = f_V(ID_A). \tag{7}$$

Digital signature of the message M by the entity A is, then, given by:

$$ID_A \| M \| E_{k_{AB_V}}(h(M)), \tag{8}$$

where $h(\cdot)$ is hash function and the common key k_{AB_V} is generated by the signer A as follows:

$$k_{AB_V} = X_A(f_V(ID_B)). \tag{9}$$

Note that the signature designates the entity B as its verifier and any other entities cannot authenticate the signature.

The verifier B authenticates the signature by the following:

1. the common key is generated by the secret-algorithm for authentication X_{B_V}:

$$k_{AB_V} = X_{B_V}(ID_A), \tag{10}$$

2. the authenticator $E_{k_{AB_V}}(h(M))$ is decrypted:

$$h(M) = D_{k_{AB_V}}(E_{k_{AB_V}}(h(M))), \tag{11}$$

3. the decrypted hash value is compared with the hash value generated again from the message.

We should note that the step 1 and step 2 is processed in TRM and that the verifier cannot forge the signature.

As already mentioned, in this scheme there are two inherent flaws that its verifier is designated and any other entities cannot authenticate the signature, and the TRM is required.

2.3   A Basic Scheme of the KPS

In this subsection, we give a brief review of *the linear scheme* for the KPS which is one of the basic schemes of the KPS. Blom[2], Blundo el al.[9], and many other schemes are regarded as extended versions of the linear scheme. After this review a new digital signature scheme which is based on the linear scheme is proposed. This means that we can also apply our digital signature scheme to many other KPSs.

In the linear scheme the center-algorithm is represented as 2nd degree co-variant symmetric tensor:

$$G_{ij} \ (i, j = 1, \ldots, m) \tag{12}$$

where each elements belong to $GF(q)$ and m corresponds to rank of G.

We, next, introduce "effective ID" which is an m-dimensional contra-variant vector transformed from ID by the one-way mapping:

$$x_A^i = f^i(ID_A) \ (i = 1, \ldots, m), \tag{13}$$

where f is one-way mapping from ID-space to the m-dimensional vector space on $GF(q)$ and it is opened to the public.

The secret-algorithm is, then, generated by contraction over $GF(q)$ of the center-algorithm and the effective ID:

$$X_{Ai} = G_{ij}x_A^j (= \sum_{j=1}^m G_{ij}x_A^j \bmod q). \tag{14}$$

The secret-algorithm is, thus, represented as covariant vector in the linear scheme.

The common key is also generated by contraction of the secret-algorithm and the effective ID or the center-algorithm and the two effective IDs:

$$k_{AB} = X_{Ai}x_B^i, \tag{15}$$
$$= X_{Bi}x_A^i, \tag{16}$$
$$= G_{ij}x_A^i x_B^j. \tag{17}$$

The common key, thus, corresponds to scalar.

The linear scheme, then, makes use of only linear operation except of ID transformation. The prominent efficiency of the linear scheme results from the above fact. Furthermore, we should note that the linear scheme achieves the lower bound of the memory size of the secret-algorithm (see, **2.1**).

3   New Digital Signature Scheme

We propose new signature scheme in this section. The new scheme is based on the quit different concept of inherence of the secret-algorithm. We introduce new cryptographic primitive "one-way homomorphism" for realization of the scheme. The one-way homomorphism Φ has the desirable property that it keeps the key-sharing ability of secret-algorithm:

$$\Phi(k_{AB}) = Y_A(ID_B) = Y_B(ID_A), \tag{18}$$

where the algorithm $Y(\cdot)$ is the transformed secret-algorithm by one-way homomorphism:

$$Y_A(\cdot) = \Phi(X_A(\cdot)), \tag{19}$$

and

$$Y_B(\cdot) = \Phi(X_B(\cdot)). \tag{20}$$

It is important to note that the secret-algorithm $X(\cdot)$ is not disclosed even though the transformed algorithm $Y(\cdot)$ may be public due to the one-wayness of Φ. We introduce a slightly modified primitive of "keyed" one-way homomorphism and claim the following statement: Mapping of the key-ed one-way homomorphism Φ with one key cannot be guessed from mappings of Φ with the other keys.

Accordingly, digital signature on the message M by the entity A is given by:

$$ID_A\|M\|S_{MA}(\cdot), \tag{21}$$

where $S_{MA}(\cdot)$ is generated by the secret-algorithm itself:

$$S_{MA}(\cdot) = \Phi_M(X_A(\cdot)). \tag{22}$$

So, the signature depends on the message and the signer but it is independent of any verifiers. This means that the signature can be authenticated by any one of the entities.

Any verifier B authenticates the signature by the following:

1. "Authenticator1: V_1" is generated from the signature:

$$V_1 = S_{MA}(ID_B), \tag{23}$$
$$= \Phi_M(X_A(ID_B)), \tag{24}$$
$$= \Phi_M(k_{AB}), \tag{25}$$

2. "Authenticator2: V_2" is generated from the verifier's secret-algorithm:

$$V_2 = \Phi_M(X_B(ID_A)), \tag{26}$$
$$= \Phi_M(k_{AB}), \tag{27}$$

3. the authenticator1 is compared with the authenticator2.

The scheme satisfies main of the digital signature requirements: 1)authenticity, 2)unforgeability, 3)non-reusability, 4)unalterability, and 5)non-repudiation.

4 Examples

The new scheme introduced in the previous section employs a magical primitive: keyed one-way homomorphism. In this section, we give two examples based on the linear scheme for the KPS and the discrete logarithm problem (DLP).

4.1 Example 1: ID-Type

Suppose that sufficiently large primes p and q are given, such that $p = 2q + 1$. We choose g with order in q on multiplicative group Z_p^*.

Mapping from $x \in Z_q$ to $y \in Z_p$ is defined by the following

$$y = g^x \bmod p, \tag{28}$$

and has one-wayness based on DLP and the following homomorphism, as well:

Addition:

$$x_1 + x_2 \bmod q \longrightarrow y_1 y_2 \bmod p, \tag{29}$$

Multiplication:

$$x_1 x_2 \bmod q \longrightarrow y_1^{x_2}, y_2^{x_1} \bmod p, \tag{30}$$

where $y_1 (= g^{x_1} \bmod p)$ and $y_2 (= g^{x_2} \bmod p)$.

Accordingly, the signature is given by:

$$S_{MA;i} = h^{X_{Ai}} \bmod p (i = 1, \ldots, m), \tag{31}$$

where h is a specific hash value of a message m and it is determined in the following way:

1. initial hash value is given by:

$$h = M, \tag{32}$$

2. calculate

$$h = h(h\|ID_A) \bmod p, \tag{33}$$

3.   – if $h^2 \neq 1 \bmod p$, stop and hash value h
    – If $h^2 = 1 \bmod p$, return to the step 1.

So, the hash value is dependent of the signer's ID. We mention it in the next section.

Authentication by the entity B is based on the following way:

1. authenticator1 is generated by:

$$V_1 = \prod_{i=1}^{m} S_{MA;i}{}^{f^i(ID_B)} \bmod p, \tag{34}$$

$$= \prod_{i=1}^{m} S_{MA;i}{}^{x_B^i} \bmod p, \tag{35}$$

$$= h^{X_{Ai} x_B^i \bmod p-1}, \tag{36}$$

$$= \begin{cases} h^{k_{AB}} \bmod p & (\text{if} \quad \tilde{k}_{AB} < q), \\ h^{k_{AB}} h^q \bmod p & (\text{if} \quad \tilde{k}_{AB} \geq q), \end{cases} \tag{37}$$

where $\tilde{k}_{AB} = X_{Ai} x_B^i \bmod 2q$. The authenticator1 is uniquely determined if the order of h is q. But the representation of V_1 has alternative ambiguity if the order of h is $2q$.

2. using the verifier's secret-algorithm, the authenticator2 is generated by:

$$V_2 = h^{k_{AB}} \bmod p, \tag{38}$$

where the common key is calculated using B's secret-algorithm:

$$k_{AB} = X_{Bi} f^i(ID_A) \bmod q, \tag{39}$$

$$= X_{Bi} x_A^i. \tag{40}$$

3. the authenticator1 is compared with the authenticator2:
In order to recognize if the signature as correct one, the following checks should be fulfilled.

$$V_1 = V_2, \tag{41}$$

or,

$$V_1 = V_2 \times h^q \bmod q, \tag{42}$$

due that V_1 has the alternative ambiguity.

4.2   Example 2: Random-Type

Suppose that sufficiently large primes p and q are given in the same way as in example 1. This time, $q|p-1$ must be satisfied. We, then, pick up g of which order is q on multiplicative group Z_p^*.

The signer A signs the message M in the following:

1. m random numbers are generated:

$$x_1, \ldots, x_m (\in Z_q), \tag{43}$$

2. the following m calculations are done:

$$r_i = g^{x_i} \bmod p (i = 1, \ldots, m), \tag{44}$$

3. using A's secret-algorithm, m pieces of s_is are calculated:

$$s_i = (h(M)x_i + X_{Ai})x^{-1} \bmod q, \tag{45}$$

where x is given by,

$$x = \sum_{i=1}^{m} x_i \bmod q, \tag{46}$$

The signature is given by (r_i, s_i).

The verifier B authenticates the signature by the following:

1. from s_i, calculate s_B:

$$s_B = s_i x_B^i \bmod q, \tag{47}$$
$$= (h(M)x_i x_B^i + X_{Ai} x_B^i)x^{-1} \bmod q, \tag{48}$$
$$= (h(M)x_i x_B^i + k_{AB})x^{-1} \bmod q, \tag{49}$$

where $x_B^i = f^i(ID_B)$ and k_{AB} is the common key between A and B,
2. from r_i, calculate r_B:

$$r_B = \prod_{i=1}^{m} r_i^{x_B^i} \bmod p. \tag{50}$$

3. from B's secret-algorithm, calculate the common key:

$$k_{AB} = X_{Bi} x_A^i \bmod q, \tag{51}$$

where $x_A^i = f^i(ID_A)$,
4. calculate the authenticator v:

$$v = r_B^{H(m)s_B^{-1}} \cdot g^{k_{AB}s_B^{-1}} \bmod p. \tag{52}$$

5. the judgment condition is given by,

$$v = \prod_{i=1}^{m} r_i \bmod p. \tag{53}$$

If the above equation is satisfied, B accepts M.

5 Security and Performance

Both examples in the previous section are based on the linear scheme and the DLP. Their security depend on information theoretic security and computational complexity.

Firstly, we concentrate on the information theoretical security. In the linear scheme, any attacker cannot disclose the center-algorithm without collecting sufficient number of secret-algorithms. This means that the linear scheme is perfectly secure if the number of its available secret-algorithms does not exceed the threshold which is equal to m in the considered examples (See **2.1** and **2.3**).

On the other hand, regarding the computational complexity it is impossible to break the system if homomorphism is exactly one-way. In the examples, in order to break the system an adversary has to solve DLPs in Eq.31, Eq.44.

The new signature scheme based on the linear scheme is also expected to fulfill such security. In order to prove this, we introduce more general one-way homomorphism instead of the special one based on DLP.

The more general one-way homomorphism is given by:

$$\Phi : (\boldsymbol{F}, +) \underset{\nleftarrow}{\overrightarrow{}} (\tilde{\boldsymbol{F}}, \oplus), \tag{54}$$

where Φ is mapping from field $\boldsymbol{F}$ to field $\tilde{\boldsymbol{F}}$ and $+$ is addition and multiplication on the field $\boldsymbol{F}$ and $\oplus$ is addition and multiplication on the field $\tilde{\boldsymbol{F}}$.

The homomorphism is, then, defined by:

$$\Phi(x + y) = \Phi(x) \oplus \Phi(y), \tag{55}$$
$$\Phi(ax) = a\Phi(x), \tag{56}$$

where $x, y, a \in \boldsymbol{F}$.

Using the one-way homomorphism, the proposal could be represented by the following.

The signature of the message M by the entity A is given by:

$$S_{MA;i} = \Phi_M(X_{Ai}), \tag{57}$$

where $\Phi_M(\cdot)$ is keyed one-way homomorphism and M is used as its key.

The authenticator V by any entity B is generated with the signature and B's ID or with B's secret-algorithm and A's ID:

$$V = \sum_{i=1,\ldots,m}^{\oplus} S_{MA;i} x_B^i, \tag{58}$$

$$= \sum_{i=1,\ldots,m}^{\oplus} \Phi_M(X_{Ai}) x_B^i, \tag{59}$$

$$= \Phi_M(X_{Ai} x_B^i), \tag{60}$$

$$= \Phi_M(X_{Bi} x_A^i), \tag{61}$$

$$= \sum_{i,j}^{\oplus} \Phi_M(G_{ij}) x_A^i x_B^i, \tag{62}$$

where $\sum$ with $\oplus$ means summation on the field $\tilde{\boldsymbol{F}}$.

These representations tell us that the signature and the authenticator are nothing but certain mappings of the secret-algorithm and the common key. From the information theory point of view, this means that any attacker cannot forge a signature without collecting m signatures of the same message. Hence, if the number of users is less than the collusion threshold, adversaris may not make any forgry of the signature.

Even if the number of users exceeds the threshold, we can avoid the collusion attack by a slight modification. In the example 1, this could be achieved by using the hash function which depends on each signer's ID, a message is hashed to be different digests for each signer. On the other hand, in the example 2 this could be achieved by using one-time random numbers so that the one-way homomorphism becomes one-time mapping. By these modifications, advarsaries are not allowed to forge a signature even if the number of colluders is more than the collusion threshold. Note that in these modifications, TRM is required. However, TRM can be also used to reduce the required memory size for the secret-algorithm.

Anyway, if the number of users is less than the collusion threshold, the new signature scheme has the same level of security as that of DLP. Assuming that it is hard enough to break TRMs, then the security is same as DLP even if the number of users exceeds the threshold.

Security of the both schemes is based on the following: 1)information theoretic security and 2)computational complexity. Finally, note that the examples require huge complexity. More precisely, their complexity is m times greater than the complexity of RSA scheme or ElGamal scheme. The size of m is usually of order 10^3 assuming the linear scheme. So complexity is tremendously huge.

6   Summary

We proposed a new digital signature scheme on ID-based key sharing infrastructure. In the new scheme, the verifier is undesignated without losing systematic robustness. To realize this, we introduced a new concept of the one-way homomorphism. Accordingly, the scheme is based on the secret-algorithm itself instead of specific common key using the one-way homomorphism. As the results, we have obtained a very sophisticated digital signature scheme using KPS infrastructure. It is shown that our digital signature scheme is secure if the discrete logarithm problem is hard enough to solve. This means that if an ID-based key sharing system has been already provided, by using it we can easily implement an ID-based digital signature system. Another remarkable property of our scheme is that TRM is not necessary, while the conventional digital signature scheme based on KPS requires it. However, in our scheme we can reduce the required memory size for the secret-algorithm by using TRM. If the security of TRM is high enough to prevent any illeagal behaviors, our scheme is realized with a considerably less amount of memory.

In comparison with other digital signature schemes, our scheme does not require any certificate of public keys since it is an ID-based signature scheme.

Furthermore, our scheme has another remarkable advantage; it provides not only the function of signature, but also key sharing.

We have proposed two implementations as examples of our digital signature scheme. The examples are also based on the linear scheme KPS. As already mentioned, the linear scheme is one of the basic scheme of the KPS, and there are many KPSs that are regarded as extended versions of the linear scheme (e.g. [2],[7]). Hence, our digital signature scheme can easily be applied to other KPSs.

References

1. A. Shamir, "Identity-Based Cryptosystems and Signature Schemes," Proc. of CRYPTO'84, Springer LNCS 196, pp.47-53, (1985).
2. R. Blom, "Non-public Key Distribution," Proc. of CRYPTO'82, Plenum Press, pp.231-236, (1983).
3. T. Matsumoto and H. Imai, "On the KEY PREDISTRIBUTION SYSTEM: A Practical Solution to the Key Distribution Problem," Proc. of CRYPTO'87, Springer LNCS 293, pp.185-193, (1987).
4. L. Gong and D. J. Wheeler, "A Matrix Key-Distribution Scheme," Journal of Cryptology, vol. 2, pp.51-59, Springer, (1993).
5. W. A. Jackson, K. M. Martin, and C. M. O'Keefe, "Multisecret Threshold Schemes," Proc. of CRYPTO'93, Springer LNCS 773, pp.126-135, (1994).
6. Y. Desmedt and V. Viswanathan, "Unconditionally Secure Dynamic Conference Key Distribution," IEEE, ISIT'98, (1998).
7. C. Blundo, A. De Santis, A. Herzberg, S. Kutten, U. Vaccaro and M. Yung, "Perfectly Secure Key Distribution for Dynamic Conferences," Proc. of CRYPTO'92, Springer LNCS 740, pp.471-486, (1993).
8. A. Fiat and M. Naor, "Broadcast Encryption," Proc. of CRYPTO'93, Springer LNCS 773, pp.480-491, (1984).
9. C. Blundo, L. A. Frota Mattos and D. S. Stinson, "Trade-offs between Communication and Storage in Unconditionally Secure Schemes for Broadcast Encryption and Interactive Key Distribution," Proc. of CRYPTO'96, Springer LNCS 1109, pp.387-400, (1996).
10. K. Kurosawa, K. Okada, and H. Saido, "New Combinatorial Bounds for Authentication Codes and Key Predistribution Schemes," Designs, Codes and Cryptography, 15, pp.87-100, (1998).
11. T. Matsumoto and H. Imai, "Applying the key predistribution systems to electronic mails and signatures," Proc. of SITA'87, pp.101-106,(1987).
12. T. Matsumoto, Y. Takashima, H. Imai, M. Sasaki, H. Yoshikawa, and S. Watanabe, "THE KPS CARD, IC Card for Cryptographic Communication Based on the Key Predistribution System," Proc. of SMART CARD 2000, IC Cards and Applications, Today and Tomorrow, Amsterdam, Oct., (1989).
13. T. Matsumoto, Y. Takashima, H. Imai, M. Sasaki, H. Yoshikawa, and S. Watanabe, "A Prototype KPS and Its Application - IC Card Based Key Sharing and Cryptographic Communication -," Trans. of IEICE Vol. E 73, No. 7, July 1990, pp.1111-1119, (1990).
14. U. Maurer and Y. Yacobi, "Non-interactive Public-Key Cryptography," Proc. of Eurocrypt'91, Springer LNCS 547, pp.498-507, (1992).
15. U. Maurer and Y. Yacobi, "A Remark on a Non-interactive Public-Key Distribution System," Proc. of Eurocrypt'92, Springer LNCS 658, pp.458-460, (1993).

16. E. Okamoto and K. Tanaka, "Identity-Based Information Security Management System for Personal Computer Networks," IEEE J. on Selected Areas in Commun., 7, 2, pp.290-294, (1989).
17. H. Tanaka, "A Realization Scheme of the Identity-Based Cryptosystems," Proc. of CRYPTO'87, Springer LNCS 293, pp.340-349, (1987).
18. S. Tsujii and J. Chao, "A New ID-Based Key Sharing System," Proc. of CRYPTO'91, Springer LNCS 576, pp.288-299, (1992).
19. Y. Desmedt and J. J. Quisquater, "Public-key systems based on the difficulty of tampering (Is there a difference between DES and RSA ?)," Proc. of CRYPTO'86, Springer LNCS 263, pp.111-117, (1986).

Cryptanalysis of Two Group Signature Schemes

Marc Joye[1,*], Seungjoo Kim[2], and Narn-Yih Lee[3]

[1] Gemplus Card International
34 rue Guynemer, 92447 Issy-les-Moulineaux Cedex, France
marc.joye@gemplus.com
[2] Korea Information Security Agency (KISA)
5th FL., Dong-A Tower, 1321-6, Seocho-Dong, Seocho-Gu, Seoul 137-070, Korea
skim@kisa.or.kr
[3] Dept of Applied Foreign Language, Nan-Tai Institute of Technology
Tainan, Taiwan 710, R.O.C.
nylee@mail.ntc.edu.tw

Abstract. Group signature schemes allow a group member to anonymously sign on group's behalf. Moreover, in case of anonymity misuse, a group authority can recover the issuer of a signature. This paper analyzes the security of two group signature schemes recently proposed by Tseng and Jan. We show that both schemes are universally forgeable, that is, anyone (not necessarily a group member) is able to produce a valid group signature on an arbitrary message, which cannot be traced by the group authority.

Keywords: Digital signatures, Group signatures, Cryptanalysis, Universal forgeries.

1 Introduction

Group signature schemes [3] allow a group member to anonymously sign on group's behalf. Moreover, in case of anonymity misuse, a group authority can recover the issuer of a signature. These schemes are especially useful in (off-line) e-cash systems [5,11] and electronic voting protocols [8] where they enable to protect user's privacy. The state-of-the-art is exemplified by the recent scheme of Camenisch and Michels [2].

Following the previous work of [9] (flawed in [6] and revised in [10]), Tseng and Jan propose an ID-based group signature scheme [12]. In [13], they also propose a group signature scheme based on the related notion of self-certified public keys [4]. In this paper, we show that their two schemes are universally forgeable, that is, anyone (not necessarily a group member) is able to produce a valid group signature on an arbitrary message, which cannot be traced by the group authority.

The rest of this paper is organized as follows. In Section 2, we review the two schemes proposed by Tseng and Jan. Next, in Section 3, we point out universal forgeries. Finally, we conclude in Section 4.

---

* Part of this work was performed while the author was with the Dept of Electrical Engineering, Tamkang University, Tamsui, Taiwan 251, R.O.C.

M. Mambo, Y. Zheng (Eds.): ISW'99, LNCS 1729, pp. 271–275, 1999.

2   Tseng-Jan Group Signature Schemes

In this section, we give a short description of the two Tseng-Jan group signature schemes and refer to the original papers [12,13] for more details.

Both schemes involve four parties: a trusted authority, the group authority, the group members, and verifiers. The *trusted authority* acts as a third helper to setup the system parameters. The *group authority* selects the group public/secret keys; he (jointly with the trusted authority) issues membership certificates to new users who wish to join the group; and, in case of disputes, opens the contentious group signatures to reveal the identity of the actual signer. Finally, *group members* anonymously sign on group's behalf using their membership certificates; and *verifiers* check the validity of the group signatures using the group public key.

2.1   ID-Based Signature

For setting up the system, a trusted authority selects two large primes $p_1 \, (\equiv 3 \bmod 8)$ and $p_2 \, (\equiv 7 \bmod 8)$ such that $(p_1 - 1)/2$ and $(p_2 - 1)/2$ are smooth, odd and co-prime [7]. Let $N = p_1 \, p_2$. The trusted authority also defines e, d, v, t satisfying $ed \equiv 1 \pmod{\varphi(N)}$ and $vt \equiv 1 \pmod{\varphi(N)}$, selects g of large order in $\mathbb{Z}_N^*$, and computes $F = g^v \bmod N$. Moreover, the group authority chooses a secret key x and computes the corresponding public key $y = F^x \bmod N$. The public parameters are (N, e, g, F, y); the secret parameters are (p_1, p_2, d, v, t, x).

When a user U_i (with identity information D_i) wants to join the group, the trusted authority computes

$$s_i = et \log_g ID_i \bmod \varphi(N) \tag{1}$$

where $ID_i = D_i$ or $2D_i$ according to $(D_i|N) = 1$ or -1, and the group authority computes

$$x_i = ID_i^{\,x} \bmod N \, . \tag{2}$$

The user membership certificate is the pair (s_i, x_i). To sign a message M, user U_i (with certificate (s_i, x_i)) chooses two random numbers r_1 and r_2 and computes $A = y^{r_1} \bmod N$, $B = y^{r_2 \, e} \bmod N$, $C = s_i + r_1 \, h(M\|A\|B) + r_2 \, e$ and $D = x_i \, y^{r_2 \, h(M\|A\|B)} \bmod N$, where $h(\cdot)$ is a publicly known hash function. The group signature on message M is given by the tuple (A, B, C, D). The validity of this signature can then be verified by checking whether

$$D^e \, A^{h(M\|A\|B)} \, B \equiv y^C \, B^{h(M\|A\|B)} \pmod{N} \, . \tag{3}$$

Finally, in case of disputes, the group authority can open the signature to recover who issued it by checking which identity ID_i satisfies $ID_i^{\,xe} \equiv D^e \, B^{-h(M\|A\|B)} \pmod{N}$.

2.2   Self-Certified Public Keys Based Signature

In the second scheme, the setup goes as follows. A trusted authority selects $N = pq$ with $p = 2p' + 1$ and $q = 2q' + 1$ where p, q, p', q' are all prime; he also selects g of order $\nu = p'q'$ and $e, d \in \mathbb{Z}_\nu^*$ satisfying $ed \equiv 1 \pmod{\nu}$. The group authority (with identity information GD) chooses a secret key x and computes $z = g^x \bmod N$. After receiving z, the trusted authority computes $y = z^{GID^{-1}} \bmod N$ where $GID = f(GD)$ for a publicly known hash function $f(\cdot)$, and the group secret key $s_G = z^{-d} \bmod N$. He sends s_G to the group authority. The public parameters are (N, e, g, y); the secret parameters are (p, q, d, x, s_G).

To join the group, a user U_i (with identity information D_i) chooses a secret key s_i, computes $z_i = g^{s_i} \bmod N$, and sends z_i to the trusted authority. The trusted authority then sends back $p_i = (z_i)^{ID_i^{-1} d} \bmod N$ where $ID_i = f(D_i)$. From p_i, the group authority computes

$$x_i = p_i^{ID_i\,x} s_G \bmod N \ . \tag{4}$$

The membership certificate of user U_i is the pair (s_i, x_i). When U_i wants to sign a message M, he chooses r_1, r_2 and r_3 at random and computes $A = r_1 s_i$, $B = r_2^{-eA} \bmod N$, $C = y^{GID\,Ar_3} \bmod N$, $D = s_i\,h(M\|A\|B\|C) + r_3\,C$ (where $h(\cdot)$ is a publicly known hash function), and $E = x_i\,r_2^{h(M\|A\|B\|C\|D)} \bmod N$. To verify the validity of signature (A, B, C, D, E) on message M, one checks whether

$$y^{GID\,A\,D} \equiv (E^{eA}\,B^{h(M\|A\|B\|C\|D)}\,y^{GID\,A})^{h(M\|A\|B\|C)}\,C^C \pmod{N} \ . \tag{5}$$

In case of disputes, the group authority opens the signature by checking which x_i satisfies the relation $(x_i)^{eA}\,B^{-h(M\|A\|B\|C\|D)} \equiv E^{eA} \pmod{N}$.

3   Cryptanalysis

3.1   On the Security of the ID-Based Scheme

To become a group member, each new user is issued a membership certificate which is then used to generate group signatures. In the ID-based scheme, the membership certificate (see Eqs (1) and (2)) is given by the pair $(s_i, x_i) = (et \log_g ID_i \bmod \varphi(N), ID_i^x \bmod N)$. So, noting that $g^{s_i} \equiv ID_i^{et} \pmod{N}$, it follows that

$$x_i^{\,e} \equiv (ID_i^{\,e})^x \equiv (g^{s_i\,t^{-1}})^x \equiv (g^{vx})^{s_i} \equiv y^{s_i} \pmod{N} \ . \tag{6}$$

Hence, if $s_i = ke$ for some integer k, then we have $x_i \equiv y^k \pmod{N}$. This means that $(s_i, x_i) = (ke, y^k \bmod N)$ is a valid membership certificate for any integer k. In particular, for $k = 0$, the pair $(s_i, x_i) = (0, 1)$ is a valid certificate. Therefore, an adversary can forge a Tseng-Jan ID-based group signature on an arbitrary message M as follows:

(1) Randomly choose r_1 and r_2;
(2) Compute $A = y^{r_1} \bmod N$, $B = y^{r_2 \, e} \bmod N$, $C = r_1 \, h(M\|A\|B) + r_2 \, e$, and $D = y^{r_2 \, h(M\|A\|B)} \bmod N$;
(3) The group signature is given by (A, B, C, D).

The above attack clearly exhibits that, contrary to what is claimed in [12], the security of the scheme does not rely on the discrete logarithm problem. Moreover, this attack can easily be generalized by considering the more general *representation problem* (see [1]) as follows. Let β be an element in $\mathbb{Z}_N^*$, and let $A = y^{a_1} \beta^{a_2} \bmod N$, $B = y^{b_1} \beta^{b_2} \bmod N$ and $D = y^{d_1} \beta^{d_2} \bmod N$ respectively denote the representations of A, B and D w.r.t. bases y and β. The verification equation (Eq. (3)) then becomes

$$(y^{d_1} \beta^{d_2})^e \, (y^{a_1} \beta^{a_2})^{h(M\|A\|B)} \, (y^{b_1} \beta^{b_2}) \equiv y^C \, (y^{b_1} \beta^{b_2})^{h(M\|A\|B)} \pmod{N},$$

which is satisfied whenever

$$\begin{cases} d_1 \, e + a_1 \, h(M\|A\|B) + b_1 \equiv C + b_1 \, h(M\|A\|B) \\ d_2 \, e + a_2 \, h(M\|A\|B) + b_2 \equiv b_2 \, h(M\|A\|B) \end{cases} \pmod{\varphi(N)} . \qquad (7)$$

The first equation in (7) is trivially satisfied if $C = d_1 e + (a_1 - b_1) \, h(M\|A\|B) + b_1$, whereas the second equation is trivially satisfied if $a_2 = b_2 = -d_2 \, e$ (over $\mathbb{Z}$). To sum up, an adversary can thus forge a Tseng-Jan ID-based group signature on an arbitrary message M as follows:

(1) Randomly choose β, a_1, b_1, d_1 and d_2, and set $a_2 = b_2 = -d_2 \, e$ (over $\mathbb{Z}$);
(2) Compute $A = y^{a_1} \beta^{a_2} \bmod N$, $B = y^{b_1} \beta^{b_2} \bmod N$ and $D = y^{d_1} \beta^{d_2} \bmod N$;
(3) Compute $C = d_1 e + (a_1 - b_1) \, h(M\|A\|B) + b_1$;
(4) The group signature is given by (A, B, C, D).

Note that the first attack corresponds to the special case $d_2 = 0$.

3.2   On the Security of the Self-Certified Public Keys Based Scheme

The self-certified public keys based scheme seems more robust. However, we will see that the previous attack still applies. From $p_i = (g^{s_i})^{ID_i^{-1}d} \bmod N$ and $s_G = (g^x)^{-d} \bmod N$, we have

$$x_i^e \equiv (p_i^{ID_i \, x} \, s_G)^e \equiv g^{s_i x} \, g^{-x} \equiv g^{x(s_i - 1)} \equiv y^{GID(s_i - 1)} \pmod{N} . \qquad (8)$$

Therefore, $(s_i, x_i) = (1 + ke, (y^{GID})^k \bmod N)$ is a valid membership certificate for any integer k. In particular, $(s_i, x_i) = (1, 1)$ is a valid certificate. As before, an adversary can thus forge a Tseng-Jan self-certified public keys based group signature on an arbitrary message M as follows:

(1) Randomly choose r_1, r_2, and r_3;
(2) Set $A = r_1$;
(3) Compute $B = r_2^{-eA} \bmod N$, $C = y^{GID \, Ar_3} \bmod N$, $D = h(M\|A\|B\|C) + r_3 \, C$, and $E = r_2^{h(M\|A\|B\|C\|D)} \bmod N$;
(4) The group signature is given by (A, B, C, D, E).

Here too, we see that the security of scheme does not rely on the representation problem (consider bases y and r_2) and a fortiori on the discrete logarithm problem. Furthermore, as before, the strategy can easily be generalized.

4 Conclusions

We have shown that the two group signature schemes proposed by Tseng and Jan are *universally* forgeable. This illustrates once more that ad-hoc constructions — although seemingly robust — certainly do not constitute a security proof and that their use always present some risks.

References

1. Stefan Brands, *An efficient off-line electronic cash system based on the representation problem*, Technical Report CS-R9323, Centrum voor Wiskunde en Informatica, April 1993.
2. Jan Camenisch and Markus Michels. A group signature scheme with improved efficiency. In *Advances in Cryptology - ASIACRYPT '98*, LNCS 1514, pp. 160–174. Springer-Verlag, 1998.
3. David Chaum and Eugène van Heijst. Group signatures. In *Advances in Cryptology - EUROCRYPT '91*, LNCS 547, pp. 257–265. Springer-Verlag, 1991.
4. Marc Girault. Self-certified public keys. In *Advances in Cryptology - EUROCRYPT '91*, LNCS 547, pp. 491–497. Springer-Verlag, 1991.
5. Anna Lysyanskaya and Zulfikar Ramzan. Group blind signatures: A scalable solution to electronic cash. In *Financial Cryptography (FC '98)*, LNCS 1465, pp. 184–197. Springer-Verlag, 1998.
6. Wenbo Mao and Chae Hoon Lim. Cryptanalysis in prime order subgroups of $\mathbb{Z}_n$. In *Advances in Cryptology - ASIACRYPT '98*, LNCS 1514, pp. 214–226. Springer-Verlag, 1998.
7. Ueli M. Maurer and Yacov Yacobi. Non-interactive public-key cryptography. In *Advances in Cryptology - EUROCRYPT '91*, LNCS 547, pp. 498–507. Springer-Verlag, 1991.
8. Toru Nakanishi, Toru Fujiwara and Hajime Watanabe. A secret voting protocol using a group signature scheme. Technical Report ISEC96-23, IEICE, September 1996.
9. Sangjoon Park, Seungjoo Kim and Dongho Won. ID-based group signature. *Electronics Letters*, 33(19):1616–1617, 1997.
10. Sangjoon Park, Seungjoo Kim and Dongho Won. On the security of ID-based group signature. *Journal of the Korean Institute of Information Security and Cryptology*, 8(3):27–37, 1998.
11. Jacques Traoré. Group signatures and their relevance to privacy-protecting off-line electronic cash systems. In *Information Security and Privacy (ACISP '99)*, LNCS 1587, pp. 228–243. Springer-Verlag, 1999.
12. Yuh-Min Tseng and Jinn-Ke Jan. A novel ID-based group signature. In T. L. Hwang and A. K. Lenstra, editors, *1998 International Computer Symposium, Workshop on Cryptology and Information Security* (Tainan, December 17–19, 1998), pp. 159–164.
13. ———. A group signature scheme using self-certified public keys. In *Ninth National Conference on Information Security* (Taichung, May 14–15, 1999), pp. 165–172.

Author Index

Lecture Notes in Computer Science

For information about Vols. 1–1649
please contact your bookseller or Springer-Verlag